ALSO BY SPENCER WELLS

*The Journey of Man*

*Deep Ancestry*

# PANDORA'S SEED

# Pandora's Seed

THE UNFORESEEN COST OF CIVILIZATION

*Spencer Wells*

*Random House / New York*

Published in the United States by Random House,
an imprint of The Random House Publishing Group,
a division of Random House, Inc., New York.

RANDOM HOUSE and colophon are registered trademarks of Random House, Inc.

LIBRARY OF CONGRESS CATALOGING-IN-PUBLICATION DATA

Wells, Spencer.
Pandora's seed : the unforeseen cost of civilization / by Spencer Wells.
p. cm.
ISBN 978-1-4000-6215-7
eBook ISBN 978-0-679-60374-0
1. Nature—Effect of human beings on. 2. Agriculture—Environmental aspects.
3. Civilization—History. I. Title.
GF75.W46 2010
304.2—dc22    2009040010

Printed in the United States of America on acid-free paper

www.atrandom.com

2  4  6  8  9  7  5  3  1

FIRST EDITION

*To Pam, for peace, love, and understanding*

*The gods presented her with a box into which each had put something harmful, and forbade her ever to open it. Then they sent her to Epimetheus, who took her gladly although Prometheus had warned him never to accept anything from Zeus. He took her, and afterward when that dangerous thing, a woman, was his, he understood how good his brother's advice had been. For Pandora, like all women, was possessed of a lively curiosity. She had to know what was in the box. One day she lifted the lid and out flew plagues innumerable, sorrow and mischief for mankind. In terror Pandora clapped the lid down, but too late. One good thing, however, was there—Hope. It was the only good the casket had held among the many evils, and it remains to this day mankind's sole comfort in misfortune.*

(AS RETOLD BY EDITH HAMILTON)

# Contents

# Foreword

*True, western societies are much better off materially than they were 40 years ago, but why is there so much crime, vandalism and graffiti? Why are divorce rates so high? Why are we seeing declines in civic engagement and trust? Why have obesity and depression reached epidemic proportions, even amongst children? Why do people call this the age of anxiety? Why do studies in most developed countries show that people are becoming unhappier?*

—RICHARD TOMKINS,
*Financial Times*, October 17, 2003

As I write this, I am 36,000 feet above the Arabian Sea, sipping a glass of wine and typing on my laptop. I'm returning from Mumbai, India, where I gave a lecture at a science and technology festival organized by the Indian Institute of Technology. I spent twice as long getting there and back as I did on the ground, and with the ten-and-a-half-hour time change I was more than a little disoriented while I was there. Still, India is one of my favorite countries and it was worth the jet-lag whiplash, bouncing halfway around the world and back over a long weekend.

The students at the institute who invited me wanted to hear about my work on genetics and human migration, a subject I have studied for the better part of the past two decades. The work my colleagues and I have carried out has shown unequivocally that all humans have a recent common origin in Africa, within the past 200,000 years, and that we only started to leave Africa to populate the rest of the world in the past 60,000. I spent an hour describing some of the new data from the field, discussing recent unpublished findings, and generally painting a picture of the genetic history of our species. At the end, as I normally do after such lectures, I took questions. These ranged from technical queries about the laboratory methods we use for analyzing human DNA to more general ones. The final question was something that I've been asked many times before, and will certainly be asked in the future: What is the broader relevance of your work?

It does seem somewhat esoteric to study the arcane details of distant human history, I suppose, but it has always fascinated me. With the right samples and a smattering of statistics, it is possible to discern the details of how our species populated the globe. "But why is this important?" the student asked. I began my answer as a scientist, describing the importance of basic research in which there is no particular practical application. Governments fund such work in many subjects, I explained, because it is possible that some new finding may end up being very important in fields that are more pragmatic—medicine, for instance. Moreover, what defines us as a species is our complex culture, and scientific inquiry for its own sake is an important part of understanding our role in the world. Imagine encountering intelligent life from another planet, I said; would such an auspicious meeting include explaining mundane details like how the latest video-game console operates or would it focus on who we are as two highly evolved species and what brought us to our present state of being? We need to learn our history to understand who we are, and to speculate on where we might be going. *"L'histoire est un grand présent, et pas seulement un passé,"* as the French philosopher Alain wrote. History is a grand view of the present, and not simply something in the past.

But there was another reason this information was important, I explained. We live today in a highly globalized world, one where people come into contact with others they may never have encountered only a century before. Africans mix with Europeans, Asians, and Native Americans to create a social mélange that is unprecedented in human history. Couple this exposure with linguistic and cultural differences, and you have a potentially volatile recipe. We are keenly attuned to such differences, and they help define how we see ourselves. Part of what our genetic work shows, though, is how trivial these differences really are—underneath the surface, at the level of our DNA, we are nearly identical. The broader relevance of our work, I explained, is that we should really start to see past the superficial features that divide us, and start to recognize that we are all part of an extended human family. To the extent that we can see ourselves as connected at the genetic level, we might be able to overcome some of our prejudices.

This notion seemed to have special relevance to the members of this audience, many of whom had recently witnessed the brutal terrorist attack in which Islamic militants from Pakistan had bombed and gunned down people in several locations in south Mumbai and taken control of two landmark hotels, the Taj and the Oberoi. Over the course of four days the terrorists killed 164 innocent people. (Nine of the militants died as well.) For India, the social impact was like 9/11 in the United States, though thankfully with fewer deaths. It would have been natural to feel enraged, to want vengeance, to use this attack as an excuse for further violence. However, the Indian reaction, according to my host at the conference, was not to dwell on negative emotions. "It has brought us all together—all India," he told me as we were weaving through traffic the night I arrived.

The nature of this whole encounter, along with the quote from Richard Tomkins that begins this foreword, highlights the theme of this book. During my career as a geneticist and anthropologist I've been lucky to work with many people around the world, ranging from senior politicians and the heads of major corporations to tribal foragers eking out a precarious existence in remote wilderness locations. What has struck me over and over again is the huge amount of change taking place in the world today, regardless of where one lives. Some of this change is good, such as the overall decrease in poverty during the course of my lifetime and the drop in the birthrate in developing countries. Other things, though, like 9/11 and the terrorism in Mumbai, have not been so welcome. Everywhere there is a feeling that the world is in flux, that we are on the brink of a historic transition, and that the world will be fundamentally changed somehow in the next few generations. The pace of technological innovation is accelerating, and we are all swept up in it. Think of all of the things indispensable to your daily life that you have learned to use only in the past decade or so. Email, Google, instant messaging, and mobile phones spring to mind immediately, but there's also hybrid car technology, curbside recycling, and social-networking sites like Facebook. All have found widespread application only since the mid-1990s, and yet today we can't imagine living without most of them. Trying to imagine what the world will be like at the close of the twenty-first century is nearly impossible.

With all of these amazing technological advances, though, has come a great deal of ancillary baggage. The unprecedented rise in chronic disease in Westernized societies is perhaps the most obvious example. I say "Westernized" rather than "Western" because of the growing incidence of heart disease, diabetes, and plain old obesity in the developing world, particularly in places such as India and China. As these societies become more like our own, they are taking on many of our worst attributes as well. Psychological disorders such as depression and anxiety are also on the rise, and drugs to treat these disorders are now the most widely prescribed in the United States. This seemingly inexorable march toward Western unhealthiness made me wonder why it happened in the first place. Is there some sort of fatal mismatch between Western culture and our biology that is making us ill? And if there is such a mismatch, how did our present culture come to dominate? Surely we are the masters of our own fate and *we* created the culture that is best suited to us, rather than our culture driving us?

The answer to this question was a long time in coming, much to the chagrin of my patient editors at Random House and Penguin. It took me on a global quest to discover the similarities between what happened thousands of years ago and what is happening now, as we face what promises to be another apparent turning point in our evolution. During the course of researching my first book, *The Journey of Man,* I was struck by the effects of the agricultural lifestyle on humans living 10,000 years ago in the Middle East. It turns out, as we'll see in the first chapter, that early farmers were actually *less* healthy than the surrounding hunter-gatherer populations. So why did the farmers "win" so resoundingly, to the extent that virtually no one on earth today lives as a hunter-gatherer? What follows is an attempt to excavate today's "archaeological" record to understand the forces that set in motion the agricultural transition and to understand how that decision created the complex world we now live in. If *The Journey of Man* was about how humans populated the world, this book is about how we have adapted both psychologically and biologically to live in it during a period of enormous change. They form two bookends to a broad view of human history that takes us from the earliest days of our species to where we might be headed as we hurtle deeper into the twenty-first century.

Stewart Brand, paraphrasing Edmund Leach in the opening sentence of the first *Whole Earth Catalog* in 1968, said it well: We are as gods; it's time we got good at it. The biggest revolution of the past 50,000 years of human history was not the advent of the Internet, the growth of the industrial age out of the seeds of the Enlightenment, or the development of modern methods of long-distance navigation. Rather, it was when a few people living in several locations around the world decided to stop gathering from the land, abiding by limits set in place by nature, and started growing their food. This decision has had more far-reaching consequences for our species than any other, and it set in motion the events that we will examine in the following chapters. With the power our species has developed as a result of these changes, we must also learn some humility. In today's world, where small groups of terrorists can inflict lasting damage on the psyche of entire nations, where apparently simple decisions can affect the biological inheritance of generations far in the future, and where more species are likely to go extinct as a result of our actions than at any point in the past 60 million years, it is time to take stock and realize that with great desires come great consequences.

PANDORA'S SEED

# Chapter One
# Mystery in the Map

*. . . the most important, most wondrous map ever produced by humankind.*

—PRESIDENT BILL CLINTON,
announcing the completion of the draft human genome sequence
on June 26, 2000

*A map is not the territory it represents.*

—ALFRED KORZYBSKI

## CHICAGO, ILLINOIS

My cab wove through the midafternoon traffic, tracing an arc along the frozen shore of Lake Michigan. On my right, the buildings of one of the world's tallest cities stabbed toward the sky, steel and glass growing out of the Illinois prairie like modern incarnations of the grass and trees that once lined the lake. A thriving metropolis of nearly three million people, Chicago boasts an airport that was once the world's busiest (it's now second), with over 190,000 passengers a day passing through its terminals—including, on this particular day, me. This sprawling city prides itself on its dynamic, forward-looking culture—the "tool maker" and "stacker of wheat," as Carl Sandburg called it. Not the most obvious place to come looking for the past.

The lake took me back in time, though—way back, before it was even there. Lake Michigan is actually a remnant of one of the largest glaciers the earth has ever seen. During the last ice age, the Laurentide ice sheet stretched from northern Canada down along the Missouri River, as far south as Indianapolis, with its eastern flank covering present-day New York and spilling into the Atlantic Ocean. When it melted, around 10,000 years ago, the water coalesced into the Great Lakes, including Michigan. Looking out the window of my cab, at the strong winds ripping across the expanse of ice reaching out from the Chicago shoreline, I felt like history might be rewinding itself. The ice age could have looked a bit like this, I thought.

This wasn't just idle musing; I've spent my life studying the past, effectively trying to rewind history. I became obsessed with it as a child, and devoured anything and everything on ancient Egypt, Greece, and Rome, the great empires of the Middle East, and the European Middle Ages. In high school biology classes I started to think about much more ancient history, its actors playing their parts on a geological stage. I added the history of life to my passion for written history, and when I got to college I decided to study the record written in our own history book—our DNA. The field I became interested in is known as population genetics, which is the study of the genetic composition of populations of living organisms, using their DNA to decipher a record of how they had changed over time. The field originated as an attempt to piece together clues about how our ancestors had moved around, how ancient populations had mixed and split off from each other, and how they had diversified over the eons. In short, *really* ancient history.

And my quest had brought me here, for the second time. My last visit to the University of Chicago—where I was headed from O'Hare—had been eighteen years earlier, in February 1989, when I was considering going there for graduate school. The lake was frozen then as well, and my early-morning walks to meetings at the university in single-digit temperatures played a small role in my decision to head to school in the somewhat warmer city of Cambridge, Massachusetts. Despite my decision, the University of Chicago was, and is, an outstanding university. Its faculty boasts brilliant researchers and thinkers in many fields, from economics to literature to physics. I had come back to visit one of them.

Jonathan Pritchard had been a graduate student at Stanford when I was a postdoctoral researcher there, and I still clearly remember his early presentations to our group. His mathematician's mind, coupled with his deep understanding of the processes of genetic change, made him a real asset to the group. We overlapped again briefly when I was at Oxford, but we lost touch over the years, although I followed his work from the papers he published in scientific journals. It was one such publication that led me to get in touch with him to discuss his findings.

This paper, published in the journal *PLoS Biology* (*PLoS* stands for *Public Library of Science,* a prestigious family of scientific journals available on the Web), described a new method his team had developed to look at selection in the human genome. Selection is the Darwinian force that has created exquisite adaptations like the eye and the ear, as well as most of the other really useful traits we humans have. As Darwin taught us, small changes that are advantageous in some way give an organism a greater chance of surviving and reproducing in the perpetual rat race that is life. Since all of these selected characteristics ultimately have their origin in the way our DNA is put together, it is logical to look to our genes to find out about what made us the way we are.

The search for selection at the genetic level has a long history, dating back to way before Watson and Crick deciphered the structure of DNA in the early 1950s. Pioneering scientists such as Theodosius Dobzhansky, a Russian immigrant to America who helped create the modern science of population genetics back in the early twentieth century, were obsessed with looking for genetic changes that could be explained only by invoking Darwin's seemingly magical force. In the days before DNA sequences could be studied directly, though, researchers observed large-scale changes in the structure of fruit fly chromosomes. (Fruit flies being the geneticist's favorite model organism, mostly because their huge salivary gland chromosomes made their patterns of genetic variation easy to study in the days before DNA sequencing.) But while they found some evidence for the past action of selection in fruit flies, the ultimate cause of the patterns they observed remained elusive.

Once it was known that DNA was the ultimate source of genetic variation, and its structure had been discovered and methods developed to determine the actual sequence of the chemical building blocks that make up the double helix (I'm glossing over about

fifty years of pioneering research here), population geneticists began to look at DNA sequences directly. In the early days (only
around twenty-five years ago), because of technical limitations,
they could examine just a few small regions in the genome (the
sum total of the genetic building blocks in an individual), and the
search for evidence of natural selection usually proved fruitless. It
was only with the completion of the Human Genome Project in
the late 1990s, and the massive technological breakthroughs that it
spawned, that scientists could finally start to reassess the issue that
had obsessed Dobzhansky and his colleagues nearly a century before: Is it possible to find evidence of selection at the DNA level
and, perhaps more interestingly, can we figure out why it has taken
place?

## GENETIC BEADS

I paid the cab driver and got out near the University of Chicago
bookstore, taking in the surroundings. Gothic-style edifices, constructed during Chicago's earlier building boom, toward the end
of the nineteenth century, surrounded me on all sides. It had been
a conscious attempt on the part of the new university—it was
founded in 1890, with funds provided by the oil baron John D.
Rockefeller—to connect with an older tradition of learning. I felt
as though I were back among the gleaming spires of Oxford, running between undergraduate tutorials. My destination, however,
was a much newer structure.

The Cummings Life Science Center was constructed in 1970; as
befitted a structure meant to house scientists engaged in the advanced study of biology, then undergoing a revolution as a result of
Watson and Crick's elucidation of the structure of DNA, the building's brick tower was bracingly modern, even a bit brutal. But I
had come to talk to Jonathan Pritchard, who was using the most
advanced techniques in genetics to look at the history of our
species. The juxtaposition of this building amid a campus of older
structures seemed fitting, given what I was here to discuss.

I located his office on one of the upper floors, and we chatted as he made me a cup of tea. An avid distance runner, with the intense, lanky look of a marathoner, he seemed somewhat surprised that I had made the trip just to talk to him. I asked him about his move from Oxford to Chicago, his personal life (one of his son's drawings hung above his desk), and what it felt like to have been granted tenure at one of the world's most prestigious universities at the precocious age of thirty-seven. He laughed, confident in his intellectual abilities, like so many of the mathematically gifted people I have known, and explained that his life was going well. We then moved on to the reason for my visit.

I wanted to talk shop. Or, rather, I wanted to get his take on the findings of his important research paper. In their *PLoS* publication, he and his colleagues had described a new method of detecting selection in the human genome. It made use of something called the HapMap, a collection of data on the so-called haplotype structure of the human genome. And to understand that we'll need to delve into the science a little.

The long string of DNA that makes up your entire genome is broken into smaller strings called chromosomes—there are twenty-three pairs of them—containing the 23,000 or so genes that direct your body to do what it does. These genes code for things like sugar-digesting enzymes in your gut, or blood-clotting proteins, or the type of earwax you have—all of the physical traits that make you who you are. The chromosomes are linear strings of DNA, composed of four chemical building blocks known as *nucleotides:* A, C, G, and T. The sequence of these nucleotides—AGCCTAGG, and so on, along the entire length of the chromosome—encodes the information in your genome and determines what each gene will do in your body. The nucleotides are arrayed along the chromosomes like beads on a string, a linear orchestra of musicians, each playing their own part in the symphony that is you. You get one of each of your chromosome pairs from your mother and one from your father.

Something funny happens to these musical beads, though, as

they are passed from your parents to you. They shuffle—like a deck of cards—partially mixing up the original linear strings of beads your parents had. That's right: your parents' chromosomes literally exchange genetic information along their lengths, breaking and reconnecting their paired strands to produce a completely new version of a chromosome to pass on to you. This is part of the reason why you don't look identical to other members of your family, but we don't know exactly why it occurs. The best theory going is that it's probably a good thing to generate novel chromosomal arrangements of the musical beads in each generation so that your child's DNA orchestra can play a different tune if times change—think about having to evolve quickly in times of intense climatic upheaval. As it's pretty much ubiquitous in animals and plants, there's almost certainly a very good reason it's there.

Probably a few readers are wondering at this point, "If the chromosomes are paired, then why does shuffling change anything? Surely they are copies of the same beads, so wouldn't shuffling them just produce the same combinations in each of the two new chromosomes?" The reason for the new combinations is that each member of a pair is actually a slightly flawed version of the other. As the chromosomes get passed down through the generations, they have to be copied by the cellular machinery for each new organism. Although this is done with great care, and there are proofreading mechanisms to make sure the copied beads look like those on the original strand, occasionally a mistake is made. By chance, one color of bead is substituted for another—a red for a green, for instance. It doesn't happen very often—perhaps a couple of times for each chromosome in every generation—but when it does happen, these changes, which geneticists call *mutations,* get passed down through the generations. They serve to introduce additional variation into the gene pool. Over time the changes have accumulated to such an extent that, on average, one in every one thousand beads differs between the chromosome pairs. Thus, each chromosome that is passed on is a shuffled version of Mom's and Dad's chromosomes, with the shuffling detectable through the patterns

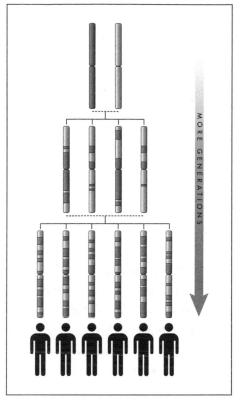

**FIGURE 1: RECOMBINATION CREATES "SHUFFLED" CHROMO-
SOMES OVER TIME.**

of the variable beads. It sounds very complicated in theory, but if
you think about it as beads on a string it is a bit easier to grasp.

What the HapMap project did was to assess the way the beads
had been shuffled in different human populations. By looking at
people from Africa, Europe, and Asia, it deduced that there was an
average length to the sections of the string of beads that hadn't
been shuffled. The length was a function of how old the population
was, the average size of the population over time, and other factors
that helped to determine exactly where on the string recombination could have occurred. The math behind all of this gets pretty
tricky, but the take-home message is that there is an average length
of these recombined places on the string of beads. Over time, many,

many generations of recombination had produced a kind of "signature" for the bead structure of a population—a pattern that served to distinguish one population's strings of beads from another's, since people living in the same geographic region tend to share more ancestors than people from different parts of the world.

Pritchard and his colleagues had developed a new statistical method to find regions of the chromosomes that seemed to have too little shuffling. In other words, they found parts of chromosomal bead strings that had long sections that seemed too similar to each other—as if everybody was wearing a uniquely patterned necklace, except that one long section of each person's necklace was pretty much identical to everyone else's. For segments like this it was possible to infer that something had happened to produce a long section of beads that seemed to be inherited like a block among many people, as though it had spread through their necklaces like a fashion accessory. One person liked the particular combination of beads they saw in part of someone else's necklace, and copied it to include in theirs. Fashion tastes served to spread the bead pattern far and wide, and pretty soon lots of people were wearing it.

Of course, chromosome patterns can't be recognized by looking at someone, and you can't just take a section of someone's chromosome and splice it into your own, so the explanation for this genetic "faddishness" had to lie somewhere else. Because the chromosomes carry genes, not beads, the inference was that the particular pattern in one person's chromosome provided some sort of an evolutionary advantage, allowing it to spread through the population. When this happens in nature, the process is known not as fashion but as selection: Darwin's force, the one that he got so excited about back in the nineteenth century, that served to create highly adapted organisms over many generations. The opposable thumb, color vision, our amazing brains—all had their origins in small changes that had been selected for in our DNA millions of years ago.

The section of the chromosomal beads that many people shared must have had a particular change in its genetic code that conferred an evolutionary advantage, and because of this the people carrying it were more likely to survive and pass on their DNA—and that popular section of beads. Studying such patterns gives us a way to peer back in time and ask how our genomes have been molded by past episodes of selection. By sifting through clues found in the DNA of people alive today, it is possible to see evidence of events that happened many generations ago, like a forensic detective trying to piece together details about a crime from evidence at the scene. And this is exactly what Pritchard and his colleagues did in their *PLoS* paper.

I asked Pritchard to explain the method, and he went to the whiteboard in his office and started to draw parallel horizontal lines in different colors, explaining how he and his colleagues had scanned the HapMap data for blocks of similar structure. He explained what they called the *integrated haplotype score*, or *iHS*, as a way of correcting for regional variation in the pattern of recombination in the human genome. The iHS accounts for the fact that some parts of the genome will have long regions that appear to be quite similar across all populations, simply due to the physical structure of the DNA in that region (and not necessarily to selection for a popular section of beads), while others will experience more recombination and will vary quite a bit among individuals. In their analysis, Pritchard and his colleagues were looking for chromosomal regions that *should* have been shuffled and diverse—and thus old—but *weren't*. These sections were younger than they should be, suggesting that something had happened relatively recently to cause a change in the pattern of that particular region of the genome—in other words, a block of similar beads had spread throughout the population due to selection.

"We were looking at events that had not gone to fixation," he explained. This meant cases where the short stretches of fashionable beads were not yet shared by everyone in the population. They did this so that they could estimate the expected level of genetic variation for each region independently, as a kind of internal control. This gave much greater power to their statistical analysis, and made it more likely that the chromosomal regions that returned a positive score really had been subject to selection.

Pritchard discussed the careful work they had done to try to account for any biases inherent in the analysis. He discussed the limitations of dealing with the three populations included in the HapMap data set, and his plans to look at data from additional populations. The HapMap was the product of an international consortium of biomedical researchers interested in finding genetic variants that could be associated with common diseases, like diabetes or hypertension. It was not planned with the primary intention of telling us more about our evolutionary history, but Pritchard's small team of researchers had discovered a way to use it to do this.

Their analysis had revealed hundreds of regions of the genome, scattered across all of our twenty-three pairs of chromosomes, that had been strongly selected. There seemed to be smoking guns everywhere, all containing stories about the evolutionary history of our species. But the most incredible single discovery to come out of the study, and the reason I was talking to Pritchard, was how recently these selective events had occurred. All of them had happened in the past 10,000 years.

We usually think of natural selection as a long, slow process. Darwin and other evolutionary biologists typically thought in terms of selection happening over millions of years, as the slight advantage provided by a new trait slowly won out over its less successful rivals. "Survival of the fittest"—the term was actually coined by the nineteenth-century social scientist Herbert Spencer—was about the gradual accumulation of genetic variants that eventually led a species to become better adapted to its environment. In studies that had been done on model experimental organisms such as bacteria or fruit flies, the calculated selective advantage was typically a fraction of a percent, which meant that organisms with the trait took thousands of generations to show evidence of selection—in other words, to show that widespread bead pattern. Pritchard's finding of hundreds of episodes of selection in the past 10,000 years—only around 350 human generations—implied that our species had been subjected to a very strong selection pressure during this time.

What could have caused this huge change in our genome?

Pritchard was quick to point out that his method may have been biased toward these results, although he admitted that his continuing analyses reinforced his finding that there had been stronger selection during this period than there had before. Other researchers have since confirmed Pritchard's results, suggesting that this time period had indeed produced a significant number of changes in our genome. To understand the timing of these events, we need to look elsewhere, outside our DNA, in the stones and bones of paleoanthropology.

## THE IMPORTANCE OF INFLECTION

Our species is a relative newcomer on the biological scene. While horseshoe crabs and sharks are recognizable in the fossil record from over 100 million years ago, the hominid lineage—composed of apes that walk upright like us—doesn't appear until around 5 million years ago. Our genus, *Homo,* appears even later, around 2.3 million years ago, with the first large-brained hominids to make stone tools, *Homo habilis,* and their descendants *Homo erectus.* Hominids with an even larger brain, looking more like us, appear around 500,000 years ago, but they still don't belong to our species. In other words, we are rank newcomers on the evolutionary scene.

According to a recent reanalysis of fossils discovered in 1967 by Richard Leakey at the Kibish Formation on the Omo River in southern Ethiopia, our species—*Homo sapiens*—first appeared as a biologically recognizable entity around 195,000 years ago. These fossils were originally dated to 130,000 years ago, but newer methods have shown them to be 65,000 years older than previously thought. In the highly contentious field of paleoanthropology, where a single discovery can rewrite history, the dates seem pretty certain. Older human remains may be discovered in the future, but as far as we know these were the first humans ever to tread the earth.

The next oldest human fossil finds date to nearly 40,000 years later, at Herto in Ethiopia and Jebel Irhoud in Morocco, but it isn't until around 120,000 years ago that significant numbers of *Homo sapiens* start to show up in the fossil record. The best-known finds are at Qafzeh and Skhul Caves in present-day Israel and the Klasies River in South Africa. The dearth of *Homo sapiens* fossils over a 75,000-year period could be due to low population densities, or perhaps it is simply a result of a poorly studied fossil record, but the implication is that humans were a relatively rare species limited to Africa and the Middle East.

Based on the number of known archaeological sites, along with clues from the human genome, we can estimate the ancient demography of our species since that time. If we plot this on a graph, an interesting pattern emerges (Figure 2). Our species had an unknown but probably fairly stable population size between the time when we originated around 200,000 years ago until around 80,000 years ago. Judging from the sparse fossil record of human remains, the population size was small and the species scattered throughout East and North Africa. Around 120,000 years ago, when we show up in the Middle East and South Africa, there was still no evidence of a significant change in the number of humans. Rather, these small, dispersed groups appear to have wandered into new territory. The Middle East at this time was basically a geographic extension of North Africa, with a similar climate, flora, and fauna, so these early humans did not venture far beyond their African home. There is no evidence of a human presence elsewhere in Asia or Europe at this time.

Between 80,000 and 50,000 years ago, however, something significant happened to the human population. The fossil and archaeological record runs dry, and there is little evidence for humans anywhere, including Africa. The settlements in the Middle East and South Africa were abandoned, as though we were retreating in the face of a catastrophic challenge. Taking the evidence at face value, the inference is that we went through a population crash. And according to recent genetic analyses, this is precisely what happened. By assessing the level of genetic diversity in the present-day human population, which turns out to be remarkably low in comparison with that of our nearest cousins, the great apes, it is possible to calculate that

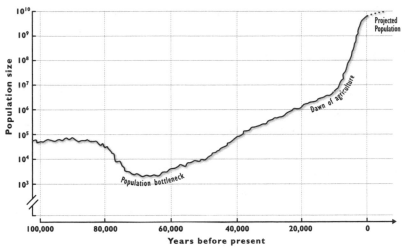

**FIGURE 2: THE VARIATION IN HUMAN POPULATION SIZE OVER THE PAST 100,000 YEARS. NOTE THE USE OF A LOGARITHMIC SCALE ON THE VERTICAL AXIS ($10^3$ = 1,000, $10^6$ = 1 MILLION, ETC.).**

the human population may have averaged no more than a mere two thousand people around 70,000 years ago. Our species was literally on the brink of extinction, at the nadir of our 200,000-year population curve. Then, around 60,000 years ago, something else happened—a change in the direction of the curve, known in mathematics as an inflection point. The human population actually started growing, and this seems to correlate with the first appearance of humans outside of Africa and the Middle East. Within 45,000 years we had spread to every continent (apart from Antarctica), increasing from the couple of thousand who survived the population crash to a few million hunter-gatherers, spread throughout the entire world. What led to this expansion will be discussed in Chapter 4.

At 10,000 years ago, something really momentous happens: we see another change in the curve, with a massive acceleration in the rate of population growth. Taking us from a few million to over six billion today, this was the true explosion of our species—the Big Bang that led to humans dominating the world stage. What set in motion this sudden growth spurt around 10,000 years ago? If you are an archaeologist, you will know the answer immediately. It was at this time that we settled down and made a conscious decision to change our relationship with nature. We developed agriculture. While hunter-gatherers had relied on *finding* their food sources, agriculturalists *created* theirs. This seemingly simple transition in the way we obtained nourishment set in motion a sea change in human history. Instead of being held captive by where we could find enough plants and animals to survive, we gained control of our food supply. The result was that food was no longer the limiting factor in determining how many people could live in one place. We'll explore how this came about in the next chapter, but for now it is enough to say that by controlling the food supply, we gained the ability to *choose* how many people could live in a particular location. If the population increased in size, it was relatively easy to grow more food. This radical change in lifestyle set in motion the Big Bang in our population growth curve.

And what's going on now? Interestingly, as we move into projections of what will happen in the twenty-first century, we see a gradual leveling off in the population curve, leading us to a steady state by the end of the century. There are many reasons for this, ranging from medical to economic, but the result will be another profound shift in our way of life. Our species has been on an accelerating growth curve since around 60,000 years ago, and for the first time since we started to expand from our ancient African homeland we will have to come to terms with life in a stagnant population. According to the United Nations, by 2050 there will be more people alive over the age of sixty than under the age of fifteen, even in much of the developing world. As we've read and heard in the news, this population shift will strain our social systems, particularly the twentieth-century concept of "retirement." It will also provide us with opportunities, as we'll see later in the book. According to Joel Cohen, a demographer at Rockefeller University, it is at this point that we will have "outgrown our childhood and adolescence as a species."

Each one of these inflection points marks a change in the fortunes of our species: our comeback from near extinction to popu-

late the world and the period of exponential growth that began
10,000 years ago—each of them has left its mark in our genes and
our culture. And in the next century we will be moving from a
rapidly expanding population to one that is more stable—or perhaps even in decline. How will this affect us, a species used to expansion and conquest?

For now, though, we're interested in the Big Bang—the massive
increase in human population that accompanied our transition to an
agricultural way of life. The date of 10,000 years ago is significant
because, according to Jonathan Pritchard's genetic results, that cor-
responds to the period in which humans have been subject to very
strong selection. We modified the plants and animals that allowed
us to develop growing agricultural societies, but judging from the
genetic data it seems that they could also have modified us.

## SIFTING THROUGH THE FALLOUT

Pritchard's results don't just point to sections of our chromosomes
that have been subject to selection over the past 10,000 years—
they also suggest which genes contained within those were the targets of selection. Typically the section where the changes have
increased in frequency contains only one, or perhaps a few, genes,
and based on their location it's possible to infer which one was the
actual target of selection. Generally, the more centrally located a
gene is within such a section, the more likely that it was the key element under selection within the chromosomal region. By compar-
ing these genes to a list of genes with known functions, another of
the spin-offs from the Human Genome Project, it's possible to
guess what function was being selected for—and therefore what
force might have been doing the selecting.

"The strongest [functional] pattern that we came up with was
for skin pigmentation," Pritchard told me as we started to discuss
the types of genes that had been selected. "There are five different
genes involved in skin pigmentation that show signals of selection
in Europeans." This helps to explain why Europeans have lighter

skin than Africans; this trait appears to have been selected for relatively recently in the European population, consistent with what anthropologists had long argued: that humans evolved originally in Africa with dark skin. It was only as we moved out of the tropics and into higher latitudes, with their lower levels of ultraviolet light, that we had to lose some of our dark pigmentation in order to allow the deeper layers of our skin to synthesize enough vitamin D—something they only do when exposed to enough UV light. The reason Europeans have pale skin—and part of the reason some of us have fair hair—is that our ancient ancestors needed to make enough vitamin D for their bones to survive the rigors of northern life thousands of years ago. I was impressed that Pritchard's genome-wide analysis had picked up this pattern without any reference to an anthropological hypothesis—he wasn't looking specifically for genes that might have been selected for skin color. I then asked him what single gene in the human genome had been most strongly selected—not a functional class of genes, like those involved in pigmentation, but the one location in the human genome that had been whipped into shape most vigorously by the action of natural selection.

"Lactase has the biggest, broadest signal," he said, turning to his computer monitor and showing me a plot of the selection patterns that's available to the general public on the Web (http://hg -wen.uchicago.edu/selection/index.html). Lactase is the enzyme that allows humans to metabolize lactose, the sugar in milk. Without it, lactose passes through our guts unmetabolized, resulting in the uncomfortable set of symptoms known as lactose intolerance. Human babies have a functioning version of the lactase gene, allowing them to survive on the milk that makes up the majority of an infant's diet, but in many human populations the gene is switched off after childhood, rendering adults unable to metabolize lactose.

Between 10,000 and 8,000 years ago, however, people living in the Middle East domesticated the goat and the cow. The animals provided a steady supply of meat on the hoof, of course, but also gave our ancestors copious quantities of milk, a nutritious, sterile

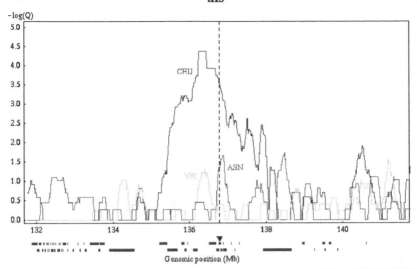

FIGURE 3: GRAPH OF THE INTEGRATED HAPLOTYPE SCORE (IHS) AROUND THE LACTASE GENE FOR THREE
HAPMAP POPULATIONS. CEU = EUROPEAN, YRI = AFRICAN (YORUBAN), ASN = ASIAN (CHINESE AND JAPAN-
ESE). NOTE THAT THE SIGNAL OF STRONG SELECTION IS VISIBLE ONLY IN THE EUROPEAN POPULATION.
SOURCE: HTTP://HAPLOTTER.UCHICAGO.EDU/.

(if collected properly) food source. It seems that in these Middle
Eastern populations, and in their descendants who brought goats
and cattle to Europe, milk was an advantageous addition to the
diet. Over time, a mutation that caused the lactase gene to remain
active after childhood increased in frequency in milk-drinking
populations. Today over 90 percent of Europeans have this genetic
variant, while the majority of Africans (apart from some cattle-
raising populations) and Asians—who never had milk as a major
component of their diets—are lactose intolerant as adults. Strong
selection for lactose tolerance in Europeans had been detected independently by other researchers investigating this unusual trait, so
it was a validation of Pritchard's analysis.

The beauty of the study was beginning to reveal itself. Instead
of focusing on individual genes with a function that could have
been useful—such as producing lactase—and trying to find evidence of past selection events, Pritchard's new technique took a
hypothesis-free approach. It simply asked where in our genome
there was evidence of selection and then tried to find genes that

could explain the statistical pattern—a shotgun approach to the study of evolution that was possible only because of the huge leaps in technology that had come about as a result of the Human Genome Project. It also revealed how much biology, and particularly genetics, was becoming a computational field, where much of the study was done in front of monitors and keyboards, rather than at the lab bench. When I'd started graduate school in genetics in the 1980s, the limiting factor in any research project had been simply generating enough data to test a hypothesis. Now data flowed like water from a fire hose, and the hard part was interpreting it and generating the many hypotheses that could explain the statistical patterns.

I asked Pritchard if there were any other interesting types of genes that showed evidence of selection, and he said yes, there were many that were involved in metabolizing food. The gene for alcohol dehydrogenase, which allows your body to break down the alcohol in that glass of wine or beer, showed evidence of strong selection, as did genes involved in metabolizing sugars and fats. He suggested that, as with the lactase gene, these would have been subject to selection as people made the transition to agriculture over the past 10,000 years. Interestingly, several genes in the cytochrome P-450 gene cluster on chromosome 1 had also been subject to strong selection. These genes are expressed in liver tissue and are involved in breaking down foreign compounds in the body, such as drugs. It is possible that new food sources could have introduced new chemicals into our diet that required new versions of these "cleansing" genes to neutralize them.

The final thing Pritchard wanted to discuss was the overlap between many of the genes that had shown evidence of selection and genes involved in complex human diseases, such as hypertension and diabetes. Hypertension, for instance, is actually a description of a symptom—high blood pressure—rather than a single disease. There are many causes of hypertension, and the complexity of this "disease" makes teasing apart genetic and lifestyle factors very difficult. One of the best-studied forms of hypertension, though, is known as "salt-sensitive" hypertension because the severity of some people's high blood pressure is strongly influenced by the amount of salt in their diet. About half of the people with high blood pressure are salt-sensitive. One of the genes that has been implicated in the salt-sensitive form of hypertension, *CYP3A,* is in the cytochrome P-450 gene cluster mentioned above—and Pritchard's analysis showed it to have been subject to strong selection in the past 10,000 years. He discussed several other examples with me as well. I asked him why something that had been selected for a *positive* function in the past would be associated with a *negative* outcome like a disease.

"The simplest explanation is that the [variant] that was favored could be protective," he explained, helping to prevent the disease in people with that particular genetic change. It makes sense—if we were in the process of adapting to an increase in dietary calories due to our new agricultural lifestyle, for instance, perhaps a variant that protected against diabetes or heart disease might be selected for. "A more complicated scenario is that things that are diseases now are a consequence of our environment. If your phenotype"— the way your genes have been expressed—"is tuned to a really harsh environment where resources are scarce, and you're trying to hoard all of your nutrients as efficiently as you can, and then the population switches to an agricultural lifestyle where you're taking in tons of nutrients, then it may be favorable to not store [nutrients] as carefully." This was an argument that dated back to the 1960s, when an American geneticist named James Neel had suggested that some noninfectious diseases in modern populations, particularly diabetes, had their roots in the transition from a hunter-gatherer lifestyle to the food-rich environment of agricultural populations, where the genes that had allowed our ancestors to store nutrients very efficiently were no longer advantageous. In other words, the genetic variant that had been good in the old environment had become bad in the new one. Pritchard noted that there are some examples, such as genes known to be involved in diabetes, where the susceptibility variant is ancestral. This is exactly what would be predicted under Neel's model.

Overall, I was getting a clear sense that the move from hunting and gathering to sedentary farming had had a significant effect on our DNA. Not only had it selected for many potentially positive changes, such as lighter skin in northern latitudes and lactase persistence in milk-drinking populations, but it had also produced some seemingly negative effects. The Big Bang in human populations had been such a rupture with our past that it left a genetic fallout that is still visible. This reminded me of another interesting fact I had learned several years ago, while researching my first book.

## WHY DID WE DO IT?

One of the great myths surrounding the development of human culture over the past 10,000 years is that things got progressively better as we moved from our hunter-gatherer existence to the sublimely elevated state in which we live today. Most people assume that the lives of our distant ancestors were, to quote Thomas Hobbes, "solitary, poor, nasty, brutish, and short." When agriculture and government came along—the inseparable duo we'll investigate in the next chapter—their obvious superiority was clear, and after that people's lives improved immeasurably. The explosion in the size of the human population after 10,000 years ago is assumed to be merely the numerical manifestation of the positive impact of growing our own food, the benefits of the new lifestyle writ in the expanding number of happy farmers. In fact, nothing could be further from the truth.

In a classic paper published in 1984, the anthropologist J. Lawrence Angel analyzed the skeletal remains of people living in the eastern Mediterranean before and after the transition to agriculture. He examined several parts of the skeletons of many individuals from each time period, focusing on teeth (which allow an estimate of the person's age at death), as well as height and something called "pelvic inlet depth index," both of which are measures of how healthy the person was. When he tabulated the data he saw a surprising pattern, shown in Table 1.

The average longevity of male Paleolithic hunter-gatherers was

## Table 1

| Historical Time Period | Pelvic Inlet Depth Index (higher is better) | Average Stature | | Median Life Span | |
|---|---|---|---|---|---|
| | | Men | Women | Men | Women |
| Paleolithic (30,000–9,000 B.C.) | 97.7 | 5'9.7" | 5'5.6" | 35.4 | 30.0 |
| Mesolithic (9,000–8,000 B.C.) | 86.3 | 5'7.9" | 5'2.9" | 33.5 | 31.3 |
| Early Neolithic (7,000–5,000 B.C.) | 76.6 | 5'6.8" | 5'1.2" | 33.6 | 29.8 |
| Late Neolithic (5,000–3,000 B.C.) | 75.6 | 5'3.5" | 5'0.7" | 33.1 | 29.2 |
| Bronze and Iron Ages (3,000–650 B.C.) | 81.0 | 5'5.5" | 5'0.7" | 37.2 | 31.1 |
| Hellenistic (300 B.C.–A.D. 120) | 86.6 | 5'7.7" | 5'1.6" | 41.9 | 38.0 |
| Medieval (A.D. 600–1000) | 85.9 | 5'6.7" | 5'1.8" | 37.7 | 31.1 |
| "Baroque" (A.D. 1400–1800) | 84.0 | 5'7.8" | 5'2.2" | 33.9 | 28.5 |
| 19th Century | 82.9 | 5'7.0" | 5'2.0" | 40.0 | 38.4 |
| Late 20th Century (USA) | 92.1 | 5'8.6" | 5'4.3" | 71.0 | 78.5 |

35.4 years, and that of females was 30.0. Women's shorter life spans were a result of complications due to childbirth, and the pattern of greater male life span has been reversed only in the past century as advances in medicine have led to healthier deliveries. Note, though, that as the population made the transition to agriculture during the Neolithic period, and particularly the Late Neolithic, when the transition was complete, the longevity for both men and women decreased significantly, to 33.1 years for men and 29.2 for women. More strikingly, the measures of health decrease dramatically. Male height drops from nearly five foot ten in the Paleolithic to approximately five-three in the Late Neolithic, and the pelvic index drops by 22 percent. People were not only dying younger, they were dying sicker. Although it is possible that this may be an artifact having to do with the particular populations studied by Angel, similar patterns have been seen in the Americas. Overall, the data shows that the transition to an agricultural lifestyle made people less healthy.

Surely for it to have resulted in such a massive expansion in human population, agriculture must have been a huge benefit to humanity. How can we explain the massive increase in human population and the dominance of agriculture, which has pretty much completely replaced hunting and gathering in every inhabited corner of the world, when it actually didn't improve people's lives? It is only in the twentieth century that we see a significant increase in longevity, and even then the pelvic index is still lower than that of our Paleolithic ancestors. Looking critically at the numbers in Table 1, an evolutionary biologist would say that hunter-gatherers had an overall 22 percent health advantage over Neolithic agriculturalists, which should have allowed them to win hands down in the game of natural selection. Why did they lose?

As we'll see in the next chapter, the story of how agriculture won this ancient competition is not a simple one. It involved an extraordinary change in our way of life, not just because it produced more people but because it marked a break from the past of a kind no other organism has ever undertaken. As hunter-gatherers, we were a species that lived in much the same way as any other, relying on the whims of nature to provide us with our food and water. When we developed agriculture we made a conscious decision to modify our environment to suit ourselves. Instead of being along for the ride, we climbed into the driver's seat. The first person to plant a seed in the Fertile Crescent 10,000 years ago set in motion events that were beyond his or her wildest imagination. Our next destination is to find out how and why those seeds took over the world.

# Chapter Two
# Growing a New Culture

*The first farmer was the first man, and all historic nobility rests on possession and use of land.*

—RALPH WALDO EMERSON,
*Society and Solitude*

*The power of population is infinitely greater than the power in the earth to produce subsistence for man.*

—THOMAS MALTHUS,
*An Essay on the Principle of Population*

## STAVANGER, NORWAY

It's worth its weight in gold," my guide shouted over the roar of the boat's engines. We were speeding along Jøsenfjord, in western Norway, on our way to see the cutting edge of agriculture. Out past the boat's wake, water and rock seemed to merge into and out of each other as though in a primeval battle. Here the land was winning, as the sheer granite of a fjord wall rocketed skyward, while elsewhere the sea claimed its supremacy, crashing over a low-lying island. With the gray skies spitting rain most of the year, and the dark winter days further blurring the distinction between land and sea, you could be forgiven for thinking that the Norwegians live an amphibious existence. Like some Atlantis in limbo, poised above and below the waterline, western Norway and its way of life are inextricably tied to the ocean. It's no wonder Viking marauders set sail from here to terrorize much of northern Europe, or that they managed to reach America hundreds of years before Columbus: the sea was in their blood.

Holding the small vial of astaxanthin in his hand, Tor Andre Giskegjerde was explaining to me how the substance is used to create the vivid pink color found in the flesh of farmed salmon. In

wild salmon, the color comes from their diet of small krill and mi-
croalgae rich in colorful carotenoids, but this culinary staple can't
be easily stored and doled out on fish farms, so an artificial ingredi-
ent is used instead. While Giskegjerde's claim about its market
value was a slight exaggeration, it was less of one than you might
think. Astaxanthin is worth so much, he told me, because the company that manufactures it still has it under patent—it's produced
from petroleum by-products via a complex chemical process. Atlantic salmon—a species once available only to anglers, accounting
for its historically high price in fish markets—is now more common
than its wild cousins, thanks to fish farming. The colorful chemical
additive fools consumers into thinking they are eating the same fish
their grandparents might have landed in a stream in Scotland, and
its cost adds nearly 25 percent to the price of farmed salmon feed.

In the icy waters of the Norwegian fjords, Marine Harvest and
its Dutch parent company, Nutreco, are engaged in research that is
changing the way we eat. (Marine Harvest has since been sold to a
Norwegian company named Pan Fish.) Of all the species that humans have domesticated over the past 10,000 years, those living in
the sea have been the most difficult to tame. In fact, it's only in the
past hundred years that we have made any headway at all. For the
most part, aquatic food production is still governed by the ebb and
flow of natural stocks. On land we may be farmers, but in the sea
we remain hunter-gatherers.

The scientists at Marine Harvest, though, have a vision of the
future of aquaculture (the growth of aquatic species for food) that
extends from salmon to fishes as diverse as cod, halibut, and tuna.
Yes, tuna: it's farmed on offshore ranches east of Australia, which
are staffed by Aussie "cowboys" who wrestle the fish into boats by
grabbing their gills and flipping them onto their backs so they are
rendered temporarily immobile. Most aquaculture is a far more
sedentary pursuit, however—and much more scientifically intensive.

The investment involved in such enterprises is enormous. Nutreco was part of British Petroleum before being spun off as an
independent company in 1994, and its staff work all over the

world. It employs dozens of scientific researchers and invests tens of millions of dollars each year in developing better strains of farmed fish. Clearly, such a venture makes sense only if there are profits to be made. And there clearly are: one of the fish pens I visited in Norway contained over $15 million worth of farmed halibut, based on the current market price—and it was a small-scale experimental facility. This is no cottage industry.

Aquaculture dates back a couple of millennia to China, where carp were held in artificial lakes after river floods. These fish, fed a diet that included silkworm feces—an early version of industrial recycling—provided a steady food resource to the growing rural population of China. Similarly, the Polynesian peoples of Hawaii constructed ponds to hold wild fish as a ready source of protein, and medieval Europeans kept carp to meet the demands for fish as the Catholic church increased the number of meat-free feast days (fish, being cold-blooded, weren't considered to be meat). These farmed carp, however, were eventually replaced by widely available wild-caught cod from the rich banks of the North Atlantic, as improved navigational and preservation techniques created one of the early modern world's most profitable international trades in agricultural products. Farmed carp were soon rendered unnecessary by a tide of wild-caught cod.

Apart from these sporadic early attempts, aquaculture is largely a phenomenon of the last century. Most of our diet—whether it's based on wheat, rice, cattle, potatoes, or any of the other non-aquatic animal and plant products humans consume—comes from species domesticated during the early days of the Neolithic Revolution, thousands of years ago. Fish and, to a lesser extent, aquatic plants have mostly been domesticated much more recently—97 percent of them since the start of the twentieth century, and nearly a quarter only in the past decade. Clearly, aquaculture is a new revolution in the making.

I had come to Stavanger to understand more about the way Marine Harvest and other companies are creating this revolution in our food supply, and to learn how and why species are domesticated in the first place. By witnessing a present-day equivalent of the Neolithic

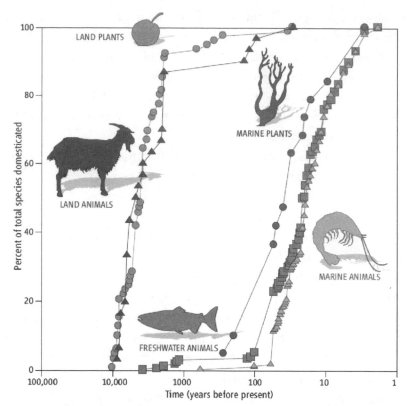

**FIGURE 4: PLOT OF DOMESTICATION RATES FOR PLANTS AND ANIMALS OVER THE PAST 10,000 YEARS. ALMOST ALL AQUATIC SPECIES HAVE BEEN DOMESTICATED IN THE PAST CENTURY. SOURCE: *SCIENCE* 316:382–83 (2007). REPRINTED WITH PERMISSION FROM AAAS.**

Revolution, I began to gain an insight into the events that occurred 10,000 years ago. The revolution in aquaculture over the past century has come about only through carefully applied scientific lessons learned from the study of the basic biology of the species involved. This is very much a high-tech enterprise, and one that makes careful use of information gleaned from fields as diverse as ecology, genetics, chemistry, and agronomy, with a huge financial investment on the part of the companies involved. It turns out that the impetus for this tremendous application of resources is not simply academic—it was born of necessity.

## EMPTY TRAPS

Twelve miles off the coast of Tunisia, an hour's ferry ride from the industrial city of Sfax, with its phosphate-processing plants and hectic medina, the Kerkennah Islands could be a world away from the twenty-first century. Like many islanders around the world, its people seem to operate in slow motion, their days governed by a slower, tidal clock. The men of Kerkennah are renowned for their fishing ability, and many in the past generation have left home to serve on boats plying the deep waters of the Mediterranean. The ones who have stayed behind, though, carry on a much older tradition.

The islands lie in shallow, grass-filled water that serves as a perfect nursery for red mullet and sea bass. It is possible to walk nearly a mile out from the beach, wading through the warm water until you feel that you could amble across the whole of the Mediterranean. The shallow depth has even meant that at various times during the ice ages, as sea levels rose and fell, the islands were attached to the African mainland. And it is the shallowness of the water that has led the people living on the Kerkennahs to develop a uniquely effective method of fishing.

Taking advantage of their knowledge of the currents running along the shore, as well as the behavior of the fish species they catch, Kerkennian fishermen use neither hooks nor nets. Rather, they trap the fish with a complex and exquisitely well-adapted system that evolved over thousands of years. Using newly cut palm fronds pushed into the sandy bottom in a tight line, they construct a long, water-permeable barrier known as a *makhloba,* stretched transversely across the prevailing currents. These fish bollards, like plane trees guiding Citroëns and Renaults along the roads of rural France, urge the fish gently into the maw of the working end. The trap itself is a series of ever-smaller chambers, constructed in a similar way from palm fronds. The first is formed by a large barrier at a sixty-degree angle from the main line, the *jerby,* which forces the

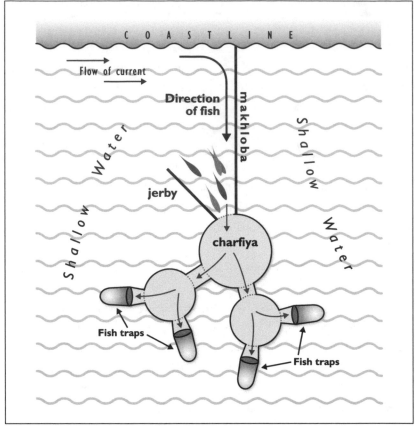

**FIGURE 5: FISH TRAP OF THE TYPE USED IN KERKENNAH, TUNISIA. THE FISH ARE GUIDED INTO THE CHARFIYA BY THE LINES OF PALM FRONDS THAT MAKE UP THE JERBY AND MAKHLOBA, AND ULTIMATELY END UP IN THE TRAPS.**

fish into a narrow opening. Once they are through this entryway, their fate is sealed; there is no escape. The fisherman call this large, round chamber the *charfiya,* or "living room," and the fish that enter find it virtually impossible to discover the entrance again. This is made particularly difficult by the construction of several other openings leading off the living room and into ever smaller chambers, the smallest culminating in an exit into a large basket. These baskets are checked and emptied twice a day by the fishermen.

While it may not be as active as other methods of fishing—particularly the intensely destructive bottom dredging used by European fishermen that is now depleting Kerkennah's stocks—it does leave the fishermen a fair amount of leisure time. Even accounting for the two weeks required to painstakingly construct a new trap, there is still plenty of time for the Kerkennians to stop and smell the roses. In these days of Kevlar-reinforced, titanium-bonded wonder materials, palm fronds are favored because their gentle decay arouses no suspicion in the fish. They need to be replaced more often, but in this case the old ways work best—the fish tend to avoid a synthetic trap. However, as the modern world, with its voracious appetite for seafood, encroaches on them, the natives of Kerkennah are finding it more and more difficult to make a living with their traditional fishing methods. Twenty years ago it was normal to recover 150 pounds of fish from a trap every day; today they are lucky to get one-tenth of that. The younger men who have shipped out on Italian fishing boats send home desperately needed cash to support their families. Others with more of an entrepreneurial bent have opened restaurants and shops catering to the new tourist industry—ironically, an industry fed by people longing to escape from the very world that threatens Kerkennah's traditional way of life.

Kerkennah's fishing methods may have been introduced to the islands by the Phoenicians, the first-millennium B.C. Mediterranean trading empire that built ancient Carthage, just north of Kerkennah where Tunisia's capital, Tunis, now stands. Similar methods are used off the west coast of Sicily to catch migrating tuna, culminating in the annual spring festival known as the *mattanza,* where the fish are corralled into smaller and smaller enclosures of nets before being killed with batons. As with the Kerkennian methods, though, the yield has dropped precipitously in the last couple of decades, and now the *mattanza* is more tourist spectacle than productive fishing technique.

Fishing is currently the only hunting activity carried out on a large scale in the modern world. While the efforts of Marine Harvest and other aquaculture companies are focused on domestication, farming still accounts for only around a quarter of all fish eaten. Most of this production takes place in China, where carp are

still raised in freshwater ponds. The majority of fish are hunted in
the sea using methods—nets, lines, and hooks—that have not
changed substantially in thousands of years. Roughly forty million Americans consider themselves to be anglers, more than
three times the number of hunters. Much has changed since we
started to grow our own food, around 10,000 years ago, and yet
fishing is an activity in which we are still very much in the preagricultural era.

    This ancient relationship is changing, though, for the same rea-
sons that the Kerkennian trappers are shifting away from their traditional methods. Overfishing has severely depleted the world's
stocks of wild fish, making it much more difficult to carry out this
ancient remnant of our hunter-gatherer past. Since the 1980s,
when fishing yields peaked, wild stocks of the major food species
have declined precipitously. The rich Grand Banks, off the coast of
Newfoundland—which may have led Europeans to North America
centuries before Columbus—have been depleted to such an extent

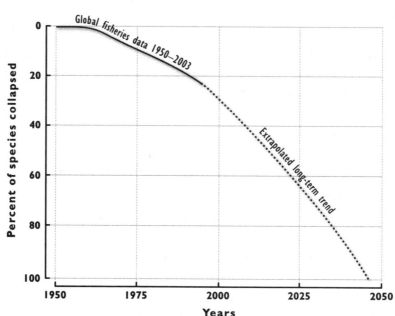

**FIGURE 6: THE CURRENT AND PROJECTED LOSS OF SEAFOOD SPECIES, WITH "COLLAPSE" DEFINED AS A DE-
CLINE IN YIELD OF 90 PERCENT OR MORE, COMPARED WITH THE LONG-TERM AVERAGE.**

that a moratorium was imposed on further cod fishing in 1995. In the North Sea off Britain, the species was declared "commercially extinct"—it no longer made financial sense to pursue it—though it has recently started to recover. According to a recent study in the journal *Science,* in 2003, 29 percent of open-sea fisheries were in a state of collapse, with a decline in yield to less than 10 percent of what they once produced.

Clearly, something is rotten in the state of the world's seas—this at a time, ironically, when people are being urged to eat fish for health reasons and fish consumption is actually increasing. The perceived health benefits of omega-3s, the long-chain fatty acids that have been shown to be important in preventing a myriad of modern ills from heart disease to senile dementia, as well as the desire for animal protein with low levels of saturated fats and the medical establishment's advocating of a lower-calorie diet, have combined to increase the consumption of fish by around a third over the past thirty years in the United States and many European countries, while beef and pork consumption have dropped. Clearly, something has to be done if the supply is to keep growing—which is where Marine Harvest and its competitors come in. By farming fish, we should be able to ensure that the supply will be virtually limitless. Whereas many early attempts at fish farming produced huge quantities of waste and led to the decline of surrounding wild stocks, modern methods are much cleaner and are better for the environment. What is clear from the data, though, is that this final vestige of our hunter-gatherer existence will soon be replaced by aquaculture—out of necessity.

Was the need to shore up a dwindling food supply also the incentive for people in the early years of the Neolithic? In an era long before notions of economic profits, why would hunter-gatherers have started farming? Particularly given the data we saw in the last chapter, suggesting that early agriculturalists lived shorter, unhealthier lives than neighboring hunter-gather populations, explaining the transition becomes much harder. What was the motive for these early agricultural revolutionaries to invest the time and resources to grow wheat and other crops while driving themselves to an earlier grave?

In *The Journey of Man,* I wrote that collective memory could have played a role. Once the seeds of agriculture were sown, it was likely that there would be no turning back from the new farming order to the old hunter-gatherer ways. "Would *you* want to make stone tools and hunt for your dinner?" I wrote. Flippant, but probably true. But there was something more: a stick to go with the newly planted carrot. These early agriculturalists probably became trapped in a catch-22 of epic proportions—one that is still playing out today.

## THE BURST DAM

The Australian archaeologist Vere Gordon Childe led quite a fascinating life. In his youth he was a Marxist who served as private secretary to the premier of the Australian state of New South Wales, as well as a talented linguist and inveterate traveler. Only later, in the 1920s, did he make the decision to pursue archaeology as a profession, first at the University of Edinburgh and later at the University of London. His early fieldwork was on Skara Brae, in the Scottish Orkney Islands, but his career really took off when he turned his attentions to the early agricultural communities of the eastern Mediterranean.

Childe coined the term "Neolithic Revolution," and he fully meant it to be taken as a revolutionary transition. All that came before was savagery (which preceded barbarity in the linear progression of his Marxist-influenced view of cultural evolution), and the fruits of civilization arose only after this momentous event. To Childe, the dawning of the Neolithic was the defining point in our history as a species, and he popularized this notion in his books, particularly his widely read *New Light on the Most Ancient East* and *Man Makes Himself,* which influenced the general public and subsequent generations of professional archaeologists.

To Childe, one of the key triggers in the onset of agriculture was the abrupt warming experienced in the Middle East at the end of the last ice age. He believed that this warming trend affected the types of plant species growing there, leading some groups of people to begin to cultivate wheat and barley. This successful experiment in cultivation led to an expansion in population and the rise of urban civilization, and gradually, from this Middle Eastern source, Neolithic farmers spread themselves (and their advanced culture) far and wide across the rest of western Eurasia.

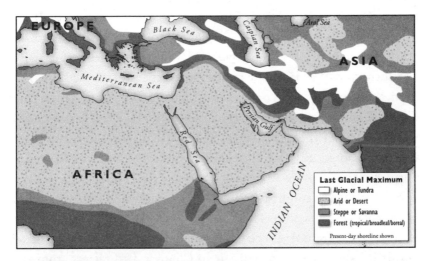

FIGURE 7: THE ECOLOGICAL ZONES OF SOUTHWESTERN ASIA AND NORTHEASTERN AFRICA DURING THE LAST GLACIAL MAXIMUM, 18,000 YEARS AGO, AND DURING THE HOLOCENE OPTIMUM, 8,000 YEARS AGO.

While basically correct, this model has since been modified with the reassessment of what happened to the climate at the end of the last ice age. Although the general trend over the past 15,000 years has been an increase in average global temperature, the period between 15,000 and 10,000 years ago was marked by abrupt advances and reversals in the warming trend. It was in this chaotic cauldron that agriculture developed.

Perhaps the best-studied part of this five-thousand-year period is a mini–ice age known as the Younger Dryas, which lasted over a thousand years, from around 12,700 to 11,500 years ago. It is named after the genus of a small plant, *Dryas octopetala,* found in the tundra regions of Scandinavia, which was replaced by forest in the southern part of its range at the end of the previous ice age but reappeared during the cooler conditions of the Younger Dryas. What led this small ice age plant to suddenly reappear is not completely clear, but the most likely theory is that its reemergence was caused, paradoxically, by the sudden melting of an ice dam in North America.

Now wait, you might be saying—it was affected by an ice dam halfway across the world? This was a very special dam, though. The warming temperatures at the end of the last ice age caused the Laurentide ice sheet we learned about in Chapter 1 to retreat from its foray into Illinois, and the large pieces of ice that remained in North America around 13,000 years ago served to confine Lake Agassiz, a massive body of fresh water located in what is today central Canada. Agassiz was composed of water from much of the rest of the former ice sheet; today's Great Lakes are impressive, but this was a monster bigger than all of them put together, larger even than the state of California or the Caspian Sea.

When the ice dam confining the lake melted, the water was released—and rather rapidly, at that. It flowed into the Saint Lawrence River basin and out into the North Atlantic. The flood of fresh water formed a kind of "shield" on the surface of the ocean— fresh water's density is lower than that of salt water—thus stymieing the Gulf Stream, which brings warmer water from the tropical

Gulf of Mexico into the North Atlantic. This natural flow had warmed western Eurasia like a massive radiator since the end of the ice age, and still does. It is the reason why palm trees grow in Cornwall, the southernmost point in Great Britain, despite the fact that its latitude is 50 degrees, the same as Winnipeg in Canada, and nearly 30 degrees north of the Tropic of Cancer. When the influx of fresh water interrupted the Gulf Stream, western Eurasia was plunged back into ice age–like conditions—the Younger Dryas.

While all of this was going on in the North Atlantic, the inhabitants of the Fertile Crescent were getting used to the warmer temperatures. Between roughly 16,000 and 12,700 years ago, the region was warming up and becoming wetter, which led to the expansion of plant species that had formerly been limited in their distribution to mountain valleys, where there were reliable supplies of water. The ready availability of these grasses—the ancestors of wheat, rye, and barley—led some human populations to focus much of their energies on gathering them, since they were a plentiful and calorie-rich food source. The Natufian people, who flourished in the western part of the Fertile Crescent during this period, were largely grain gatherers. And unlike almost all hunter-gatherers who came before, they were sedentary—they lived in small villages.

Since the earliest days of hominid evolution, our ancestors had been seminomadic. This is because of the uncertainty inherent in being a hunter-gatherer—if the food supply dwindles in one place, you pack up and move to better foraging grounds. It was our ability to do this successfully that led our ancestors out of their homeland in Africa again and again over the past few million years. *Homo erectus,* who left around 1.8 million years ago, was simply following food, as was *Homo heidelbergensis,* who left around 500,000 years ago and gave rise to the Neanderthals in Europe. We are the descendants of a third wave of African hunter-gatherer migrants that left around 50,000 to 60,000 years ago, as I detailed in *The Journey of Man.* And throughout this long line of human evolution we were consistently seminomadic, staying in one place only as long as the pickings were good and moving on when they weren't.

This started to change toward the end of the Paleolithic period—the period that preceded the Neolithic—when all humans were hunter-gatherers. According to recent research carried out in Israel by archaeologist Dani Nadel and his colleagues, there is evidence for grain gathering and flour making at one of his sites near the Sea of Galilee dating back to 23,000 years ago. This would place that community within the early part of the time period allocated to the Kebaran culture, which preceded the Natufian in the Levant. The Kebarans were the link with the true hunter-gatherers of Middle Eastern prehistory—highly nomadic, shifting their settlements seasonally to follow the food and water supplies during the cold, dry terminal period of the last ice age. But even these mobile hunter-gatherers seem to have recognized the advantages of collecting and grinding wheat, albeit in much smaller quantities than the Natufians.

When the last ice age ended, though, wheat expanded its range. Life became easier for our ancestors; food that had been difficult to obtain during the cold, dry conditions of the last ice age suddenly became more plentiful. This allowed these people to finally settle down in an area with large quantities of the easily gathered grain. Gathering wild wheat yields more calories of food for each calorie of energy invested than did early forms of agriculture, which made this grain a fabulously valuable food source for the Natufians. Moreover, it was a particular *type* of food resource that lent itself to long-term storage—a seed that could be stored dry for years. A couple of weeks of intensive grain gathering in the fall could yield enough wheat to feed a family for a year, if supplemented with nuts and game meat. Life was good, and they made the most of it by . . . well . . . doing what people do. They had babies.

The hunter-gatherer way of life had limited the number of children people had as part of a complex feedback loop with the environment. If the population grew too large it was necessary to split and form two smaller groups, one of which would typically move on to new hunting grounds. The calorie-rich environment of the

Fertile Crescent wild grain fields increased the region's carrying capacity (the number of people the land could support), and the human population responded. Natufian settlements during this period expanded into villages of 150 people or more, complete with circular houses and storage pits. It was a radical shift in our relationship with nature, and it happened only because the Natufians could rely on a steady supply of grain from the territory where they lived.

Then, suddenly, it all changed. That burst dam thousands of miles away in North America, setting in motion the Younger Dryas, brought a return of the long winter. The population of the Middle East was cast back into the ice age, but this time it had a strike against it: the people couldn't move on to greener pastures. They had invested too much in their villages, the collective memory of the good times was probably still fresh in their minds (leave the village to return to the hardscrabble life of a nomad? Unthinkable!), and in all likelihood there were now too many people to return to life as nomadic hunter-gatherers. The Natufians were in a bind.

Although *Dry*as refers to a cold-tolerant plant species, it could perhaps more aptly refer to the *dry*ing effect during these periods of global cooling, for this was the main result in the Middle East. As the land dried out, the wild grain retreated from the lowlands, remaining only in the higher mountain valleys, where it could get enough water. The Natufians had to travel farther and farther from their lowland settlements to gather enough to survive. This would have put tremendous pressure on the food supply, and probably resulted in an increased mortality rate in these people accustomed to a land of plenty. It was humanity's first real encounter with Thomas Malthus's conjecture that population growth will eventually produce more people than can be supported by the available food supply.

Around this time we also see evidence for the extinction of megafauna, large mammals, such as the woolly mammoth and Irish elk in Europe, that would have formed part of the diet of these hunter-gatherers. While such extinction events had happened before, most notably in Australia when humans first arrived there

around 50,000 years ago, as well as in North America with humanity's arrival around 15,000 years ago, the clearest evidence of their
occurrence in Europe and the Middle East comes at the end of the
last ice age. This extinction event is further evidence that climate
change and human population pressures were having a significant
effect on food resources. A population that had been able to live sus-
tainably during the warm period immediately following the end of
the last ice age was now too large to be supported by the diminished
resources of the Younger Dryas, and the animal species lost out.

Then, sometime between 12,000 and 11,000 years ago, one of
these stressed Natufians had a revolutionary idea. What if, instead
of walking farther each day to gather food, they simply planted it
close to the village? It was probably a woman, since women typically did the gathering in hunter-gatherer populations and thus
had access to the seeds—and an incentive to reduce their gathering
commute! Her first efforts must have been rewarded with admira-
tion from the entire village, and the idea quickly spread. Virtually
overnight humans had gone from being controlled *by* their food
supply to controlling *it*.

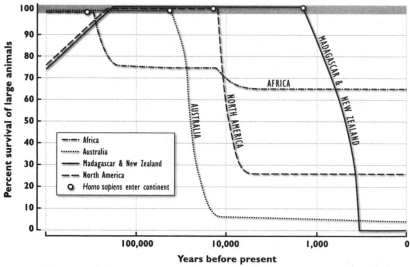

FIGURE 8: PATTERN OF MEGAFAUNA EXTINCTIONS ON THREE CONTINENTS, MADAGASCAR, AND NEW
ZEALAND. IN EACH CASE THE EXTINCTION OCCURRED SOON AFTER THE ARRIVAL OF HUMANS.

This course of events can actually be seen in the bones of the people who lived in the region at the time, making use of something called the strontium/calcium ratio. Strontium (Sr) is an element that accumulates in human bone, its level determined by its abundance in the groundwater of the region (as well as exposure to nuclear fallout, which has Sr-90 as one of its major constituents). Plants absorb strontium from the water as they grow, then pass it on to the animals that eat them. Thus, the higher the proportion of plant food in your diet, the higher your strontium levels. Natufian remains have a very high level of strontium during their intense gathering phase prior to the Younger Dryas; the level drops significantly as the wild stands of grain shrank and they turned to hunting to survive. The level then rises dramatically after the onset of domestication as the new culture took hold and plants made up a larger proportion of their diet.

This change in our relationship with nature had an extraordinarily far-reaching impact on the future of humanity—it was about much more than just food. We'll examine the early fallout from cultivation a bit later in this chapter, but for now we need to zoom out from our close focus on this one region. While this sequence of events was playing out in the Middle East, extraordinary things were happening in other places around the world. The Natufians were not alone in their early attempts to domesticate crops; cultivation seems to have been a global trend around this time. But in the days before mass media and the Internet, how did the seed of this revolutionary idea get planted in places as far afield as Mesoamerica, southern China, and New Guinea? And what does it reveal about the next stage in the development of agriculture—the crucial step from planting wild seeds to their domestication, coupled with selective breeding for desirable traits?

## PEAKS AND VALLEYS

The Soviet botanist and geneticist Nikolai Vavilov led one of those lives that deserve to be a feature film. Born into a bourgeois merchant family in Moscow in 1887, he spent several years during his youth traveling and studying in Europe, returning home just in time for the Bolshevik Revolution. He became a prominent member of the Soviet regime, a member of the Supreme Soviet, and a recipient of the prestigious Lenin Prize. He created the Institute for Plant Industry in Leningrad (now Saint Petersburg), and headed it for nearly twenty years; today this institute and the Institute of General Genetics in Moscow both bear his name. Yet due to the bizarre rise of an agronomist named Trofim Lysenko in the 1930s and his pseudoscientific attacks on genetics and the basic rules of evolution (both championed by Vavilov), in 1943 this great scientist starved to death in a gulag, having been jailed by Stalin in 1940 for allegedly plotting to destroy Soviet agriculture.

Vavilov was a hugely influential thinker on the origins of plant domestication, and the institute in Saint Petersburg still houses one of the world's largest seed banks, created in an effort to preserve and study the diversity of cultivated crops from around the world. During the twenty-eight-month Siege of Leningrad, the workers managed to protect the collection from the city's starving residents, who tried repeatedly to eat its contents, and it is still a major botanical resource to this day.

In his extraordinarily influential work on domesticated plants, Vavilov described many primary centers of plant domestication. One was the Fertile Crescent, which we've just learned about. Other major centers were in China, Mesoamerica, and the Andes of South America. A wide range of places, but all are similar in one way: they are all mountainous regions. Why not coastal areas or prairies? Primarily because mountains serve as so-called *refugia* of biological diversity—places where species continue to thrive when the surrounding plains are too dry to sustain them, due to climatic shifts such as those that have occurred frequently throughout the past few million years. Because mountains draw more rainfall, they serve as relatively safe havens in times of climatic stress, so they are the places where genetic diversity is typically the highest. And high genetic diversity allows for the development of advantageous traits that can be selected for by humans, including seed retention (as opposed to a plant's jettisoning its seeds when they are mature) and other characteristics that suit species' use as food crops.

Humans can't live easily in high mountains—we tend to prefer lowlands, for climatic reasons—but plants advance and retreat, "breathing" in and out of the lowlands during wetter and drier phases. This provides us our first clue as to why domestication happened in all of these places at the same time.

Mesoamerica, for instance, has given us many crops that are indispensable components of the modern diet: corn, tomatoes, beans, chilies, chocolate, vanilla, squash, pineapples, avocados, and pumpkins. Many were domesticated in the region of present-day Oaxaca, in southern Mexico, which has a rugged, mountainous terrain that has served to fragment human populations, resulting in a tremendous amount of cultural and linguistic diversity to match its botanical horn of plenty. Corn is far and away the most important Oaxacan crop, and evidence shows that it has been cultivated since around 10,000 years ago. There is some debate about corn's botanical ancestor—its closest wild relative, teosinte (pronounced tay-o-SIN-tee), is so different in form that many scientists find it difficult to believe that one developed from the other—but not about its geographic origin.

Domesticated corn later spread far from its Mexican homeland, reaching into North and South America over the subsequent 8,000 years, much as wheat and barley spread far from their origin in the Fertile Crescent. The spread of corn has been well documented from human remains in North America, where the sudden transition from hunting and gathering can be seen in the "carbon signature" in the bones, in a similar way to that in which strontium revealed the sequence of Neolithic events in the Middle East. This is because hunter-gatherers eat primarily what are known as $C_3$ plants, which use carbon dioxide from the atmosphere to produce molecules with three carbon atoms as their energy store. Some 95 percent of the world's plants are members of this $C_3$ group, and it was the first to evolve, over 250 million years ago. A more efficient type of plant metabolism, known as $C_4$, evolved more recently, within the past 65 million years. The C4 plants include mostly tropical grasses, such as corn, millet, and sugarcane, that store their energy in 4-carbon molecules.

There is one other difference between C3 and C4 plants, and this is how this little foray into plant physiology fits into our story. The carbon atoms in the atmosphere—coming from your breath or the car exhaust from your morning commute—that these plants use to make sugars and starches aren't all identical. There are several different variants of carbon, distinguished by their atomic anatomy. Most carbon molecules have 6 protons and 6 neutrons packed into each nucleus, for an atomic weight of 12 (6 + 6). However, rarer forms of carbon have 7 or even 8 neutrons packed in with their 6 protons, giving them atomic weights of 13 and 14. Carbon-14 is extremely rare, but its tendency to lose atomic baggage (an electron and an antineutrino, if you must know) in an effort to drop a bit of extra molecular weight makes it extremely useful as a way of dating once-living material. Carbon-13 doesn't decay; it sticks around indefinitely and gives us another tool in our archaeological atomic arsenal. It turns out that C3 plants, for whatever reason, are picky and don't like carbon-13, excluding it from their metabolic machinery. C4 plants don't seem to care and will use whatever is available. This means that C4 plants have higher ratios of carbon-13 to carbon-12 than do C3 plants.

So what does all of this mucking about in the world of carbon atoms mean? When people add C4 plants (like corn) to their diet, the ratio of carbon-13 in *them* also increases. By carefully measuring these carbon ratios in ancient bones, we can see when people started to eat C4 plants like corn. And when we do this in North America, examining bones from around the time that corn started to spread into a region, we can see a dramatic increase. While the ratios aren't necessarily indicative of the actual amounts in a person's diet (it's unlikely that 75 percent of their calories came from corn, as Figure 9 might suggest), it does indicate the extraordinary shift in diet that accompanied the spread of agricultural "killer apps" like corn and wheat.

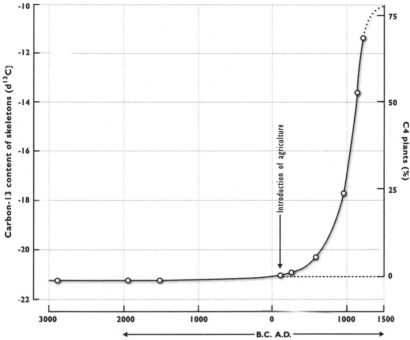

FIGURE 9: THE SPREAD OF CORN INTO NORTH AMERICA, AS CHRONICLED IN ARCHAEOLOGICAL SITES FROM THE MIDWESTERN UNITED STATES. PRIOR TO THE INTRODUCTION OF AGRICULTURE, THE CARBON ISOTOPE SIGNATURE OF HUMAN SKELETONS IS FLAT, BUT IT INCREASES DRAMATICALLY AFTERWARD, A RESULT OF LARGE AMOUNTS OF THE C4 PLANT CORN ENTERING THE DIET.

Similarly, rice seems to have been domesticated first in the mountains of southern China and northern India, where its wild ancestor *Oryza rufipogon* still grows. Through careful analyses of phytoliths, microscopic stonelike particles in plants that serve as a kind of species fingerprint and are preserved in the archaeological record, Zhijun Zhao of the Smithsonian Tropical Research Institute has found evidence that hunter-gatherers living on the Yangtze River in central China were eating rice around 13,000 years ago. With the onset of the Younger Dryas and the cooler temperatures in the Northern Hemisphere, however, the rice phytoliths disappear from the archaeological record, reappearing only around 11,000 years ago, when warmer and wetter conditions returned—and judging from changes in the phytoliths, these appear to have been cultivated. It seems that during the Younger Dryas the rice retreated back to a more hospitable environment, and humans—as in the Middle East and the mountain valleys of Oaxaca—were forced to start planting it to keep the grain in their diet.

Thus, in the centers of domestication for the three main grain crops around the world, we see a similar interaction between hunter-gatherers and their local grains. Intensive foraging at the end of the last ice age, coupled with warmer, wetter conditions, led to specialized gathering of particular plant species and an increase in population. The onset of the Younger Dryas created a crisis in food supply, which forced these sedentary foragers to start cultivating grains that had previously been plentiful in the wild. The combination of a demographic expansion followed by a climatic stress probably explains why we see the development of agriculture independently at the same time around the world. Cultivating food allowed these populations to survive the cold snap of the Younger Dryas, and when favorable conditions returned, agriculture was ready to take off. All that was needed was one final step: domestication.

## THE IMPORTANCE OF DUPLICATION

Today Captain William Bligh's name is synonymous with cruel leadership, but in fact he was quite a good naval commander. His fall to ignominy came from his tough treatment of his crew during a six-month sojourn in Tahiti in 1789, during which he tried to enforce a ban on "liaisons" with the local women. Ironically, he was motivated in large part by his concern for the Tahitians; Bligh didn't want his crew to spread sexually transmitted diseases on the island. If they hadn't spent so much time in Tahiti, the *Bounty* would probably be a footnote in naval history textbooks and history would have a different term for a cruel authoritarian. Unfortunately, Bligh couldn't leave Tahiti any sooner because of the difficulty of cultivating breadfruit.

Unlike most plants humans eat, breadfruit typically has no seeds (there are a few varieties that do, but they aren't widely cultivated). The only way to propagate the plant is by air layering, a tedious process in which a small incision is made on a branch or stem and then wrapped with a rooting medium, like moss or soil. After a few weeks new roots will have grown from the incision and the branch can be removed and planted on its own, eventually yielding a new, independent plant. Two botanists from the Royal Botanic Gardens at Kew accompanied Bligh on his voyage in order to perform this time-consuming task. It was the only reason for the *Bounty*'s multiyear journey—Bligh was supposed to deliver as many of the plants as possible to British colonies in the West Indies in order to feed the burgeoning slave population there. Breadfruit, despite the tedious business of propagation, is a calorie-rich and easily grown food source.

The complicated techniques that allow breadfruit to be cultivated are among the many that have been developed by humans since the dawn of agriculture. Propagation is one of the key parts of domestication, because without it you can't make more of your food source. Doing it consistently requires a tremendous amount of knowledge about the biology of the species in question—life history, preferred growth conditions, and many other details. Most traditional peoples acquire this knowledge through observation and trial and error, ultimately passing it on through word of mouth; modern research is being applied to the same end in the more recent development of aquaculture.

During my conversations in Stavanger with Marine Harvest's researchers, I heard two words repeated by many different people: "closed cycle." In farm-speak this refers to the ability to breed animals and plants in captivity so that there is no input from the wild—the breeding cycle is "closed" and you can grow as many of the species in question as you want. This is the real test of a farming operation, and the true meaning of domestication. It is also what much of Marine Harvest's research-and-development budget is spent on: finding a way to coax species through the various stages of birth, growth, and reproduction. While we take it for granted that ranchers can breed cattle and poultry farmers can raise chickens from eggs to adulthood, it actually required a great deal of effort to figure out how to do it in the first place—as with the propagation of breadfruit.

Perhaps the best-studied example of the development of agriculture during the Neolithic period is described in an epic book about the excavation of a place called Abu Hureyra, in northern Syria. In *Village on the Euphrates,* Andrew Moore and his coauthors laid out the critical role of propagation early on: "Domestication may be defined in several ways . . . but the essence of it is that humans usually influence the breeding of the species concerned." This is the key step in creating more of the species; if your animals and plants don't produce offspring, you have to keep going back to the wild for more.

Domestication is about far more than simply making more of the species in question, though. It is also about selecting for traits that make that species a better source of food. For instance, wild cod typically mature in four to six years, but Marine Harvest's scientists have selectively bred fast-maturing individuals who reach adulthood in only two. Shorter time to maturity means more fish in the markets—it makes sense. Domestication is also about modifying the farming environment to make sure the species in question has what it needs to thrive. Halibut, for example, are currently being studied by Marine Harvest scientists to see if they are a viable species for domestication. While they yield delicious flesh that is very popular on European and American dining tables, their unusual biology has forced the scientists to get creative in order to raise them. This is because, although halibut are born looking like normal fish, after about six months one of their eyes literally migrates along the top of the head, so that a few months later both eyes are on the same side. The fish starts to swim on its side and becomes a bottom dweller, preferring to live next to a horizontal surface. This aspect of its life history is problematic for fish farmers, as it severely limits the number of fish that can be raised per unit area—the third dimension in the pen isn't really being used by the halibut, because they all want to be on the bottom. To get around this problem, Marine Harvest's scientists have devised "halibut high-rises" composed of columns of mesh shelves that allow more fish to be packed into the same diameter pen. The strategy seems to

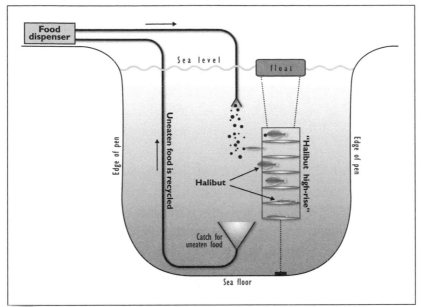

**FIGURE 10: A "HALIBUT HIGH-RISE."**

be working: the fish happily swim out from the high-rise to eat food that falls from the surface, then return to their perches to rest.

Selecting for the traits that allow for domestication is easier in some species than in others, and it turns out that our Big Three—wheat, rice, and corn, which together provide more than half of all calories consumed in the modern world—are particularly suitable for selective breeding. All of them are what is known as *polyploid,* which means their genomes have been duplicated—the chromosome number literally doubled at some point in the past few million years, in some cases more than once. Many plants are polyploid, and genome duplication seems to have happened quite often during plant evolution. It's as if you photocopied the entire genome and inserted that spare copy into the nucleus of the cell. This has some pretty interesting consequences for what you can do to the plant with selective breeding.

When you have a spare copy of something, you can take more risks than if you had only one. It's kind of like the "lives" you have in a video game—you can make poor choices and still keep play-

ing. This holds true at the genetic level as well, since having a duplicate allows you to tinker with one of the copies while retaining
an unaltered version. It gives you a backup, in other words, in case
something goes wrong in your tinkering. Duplicate copies can
open up new opportunities for evolutionary change without risk-
ing the loss of vital functions—and can sometimes lead to more
rapid evolution. This idea was first championed by Susumu Ohno,
a Japanese-American population geneticist, who wrote a classic
book entitled *Evolution by Gene Duplication* in 1970. In this book
Ohno presented what he believed to be one of the fundamental
mechanisms of molecular evolution: duplicated genes leading to
rapid evolutionary change due to the relaxed selection made possible by having a backup copy. He also coined the term "junk
DNA," referring to the large stretches of DNA in the genome with
no known function. This is the ultimate fate of gene duplicates
that suffer a fatal mutation and become nonfunctional. But the
working copy of the gene keeps the organism alive. Almost all cancerous tumors duplicate their DNA, becoming polyploid as they
develop. Geneticists believe this duplication of key genes gives
them more plasticity—more options to develop in ways that normal cells never would.

In addition to being polyploid, wheat, rice, and corn also appear
to have a very high rate of mutation. Their DNA, it seems, is in a
constant state of flux, duplicating and deleting parts in a molecular
shuffle that produces a high level of natural variation in many traits.
Some of this shuffling is caused by the presence of small DNA parasites known as transposable elements. These are like little viruses
embedded in the genome, and may be the remnants of what were
once active retroviruses (a family that includes HIV, the human immunodeficiency virus) that lost their ability to infect other cells but
retained the retrovirus's penchant for integrating into DNA and
hopping around. Their discovery by Barbara McClintock in the
1940s and 1950s, during her efforts to understand some of the odd
characteristics of corn genetics, was initially met with skepticism
from fellow geneticists, but later research showed her work to have
been correct, and she was awarded a Nobel Prize in 1983.

Corn is a wonderful example of how careful selective breeding produced characteristics that are a far cry from those of corn's likely wild ancestor. Teosinte looks completely different from today's cultivated corn—as you can see in Figure 11—and while it is impossible to calculate the level of observational research and selective breeding required to domesticate wheat, rice, and corn during Neolithic times, it must have taken an enormous amount of effort to select for such extreme changes. In fact, recent genetic studies have suggested just how difficult it must have been.

Three genes, known as *teosinte branched 1* (*tb1*), *pro-lamin box binding factor* (*pbf*), and *sugary 1* (*su1*), are key to creating certain traits that distinguish corn from teosinte. Despite their complicated names and even more complicated biochemical functions (*tb1*, for example, determines how the cobs are arranged on the corn plant, while *su1* determines the mix of sugars found in the corn kernel), all seem to have been under strong selection as early as 4,400 years ago, according to a recent analysis of these genes in ancient corn remains. However, selection for these traits seems to have been ongoing nearly 2,000 years later, showing how difficult the process of selection is, particularly in a society lacking today's scientific knowledge. That these early Mexican farmers were able to create corn from teosinte is remarkable.

The genetic plasticity of wheat, rice, and corn gave them an edge over other potential food crops, and it is a large part of the reason that they are so widely cultivated today. The Natufians at Abu Hureyra consumed around 150 different plants, gathered (along with wheat) from the rich hills of the northern Fertile Crescent, but by the time domestication was complete, a few thousand years later, their diet had dropped to only eight species, and wheat was by far the most important dietary component. Today, the Big Three cereals account for around 90 percent of grain species under cultivation—they won the race to be humanity's most important foods.

But here is the sting in the tail. When we used our amazing ingenuity to select for traits that allowed us to cultivate these incredibly successful foods, we unknowingly set in motion a strong selective regime on ourselves. In some ways their success has cre-

FIGURE 11: TEOSINTE (LEFT) AND CULTIVATED CORN, SHOW-
ING THE PROFOUND DIFFERENCES BETWEEN THE TWO.
(PHOTO COURTESY OF JOHN DOEBLEY.)

ated a sort of nutritional "cancer" that has taken on a life of its own,
coming to dominate our diet in ways those first farmers probably
never could have imagined. (Michael Pollan's excellent book *The
Omnivore's Dilemma* discusses the rise of modern corn agriculture in
a chilling overview of industrial farming.) The end result of the
higher—and less complex, in terms of species mix—carbohydrate
levels in our diet is still playing out, as we'll see in the next chapter. But first, back to the immediate fallout from agriculture: more
people, and what to do with them.

## SOCIAL MALIGNANCY

We now come to the crux of what this book is all about. When our
ancestors created agriculture around 10,000 years ago, they had no
idea of what other changes they were setting in motion. They were
simply responding to an immediate need for more reliable sources
of food during a time of climatic stress, obviously making decisions
about the future based on the near term rather than how events
might ultimately play out. They were unaware of what, by chang-

ing their fundamental relationship with nature, they were unleashing on the world. Instead of relying on nature's plenty, they were creating it for themselves. By doing so they divorced themselves—and us—from millions of years of evolutionary history, charting a new course into the future without a map to guide them through the pitfalls that would appear over the subsequent ten millennia.

The first unintended consequence of this change was that more food produced more people. While the human population had been growing slowly since around 60,000 years ago as we slowly expanded around the world as hunter-gatherers, this was an entirely different form of growth because it was both *faster* and *geographically constrained.* Humans had been held back in their population expansion by the ice age, when the challenging climatic conditions made life more difficult. If, as genetic evidence suggests, we started with a population of a few thousand people living in Africa around 70,000 years ago, by the beginning of the Neolithic period our number had grown to only a few million—a thousandfold increase in 60,000 years. With the invention of agriculture, though, the growth rate suddenly jumped to levels never before seen in the evolutionary history of a large primate. The small Natufian villages became towns virtually overnight, and this brought the second consequence of agriculture: the development of governments.

In a typical hunter-gatherer society, most people are of roughly equal social status. Of course there is variation in skill levels—one person may be better at hunting, another better at telling stories that pass on the lore of the culture, another perhaps the best at sewing—but all are recognized as contributing to the society. Disputes are generally settled in one of two ways, either through group adjudication or tribal fission. Fission, however, is impossible when there is limited access to key resources, such as cultivable land or water supplies. It becomes impossible for part of the tribe to simply move to a new place, since the resources won't necessarily be available elsewhere; in addition, sedentary people have likely invested a large amount of effort in creating their houses and other immobile material goods and will be reluctant to leave them. The need to remain in a particular place led certain rare hunter-gatherer communities, such as the pre-Columbian Native American communities of the northwest coast of North America and the Paleolithic Jomon population of Japan, to create their own complex systems of government. The large population sizes that were reached through the exploitation of incredibly rich fishing grounds, and the necessity of preserving access to these areas, meant that these communities faced similar pressures toward organization as did early agricultural communities. These coastal dwellers are generally recognized as special cases, however, and the vast majority of hunter-gatherers surviving today have—as those in the past likely had—a more egalitarian social structure.

What were these early forms of Neolithic government like? It's probable that they started out relatively egalitarian, as suggested by archaeological sites such as Çatalhöyük, in present-day Turkey. This Neolithic site dates from around 9,500 years ago, and over the six millennia it was inhabited, it probably fluctuated in population between five thousand and ten thousand people. Clearly a departure from the small Natufian and early Neolithic farming villages, it was the world's first city, according to its discoverer and excavator, James Mellaart. However, despite its considerable population, the buildings are all roughly comparable in size and appear to have been used as dwellings. The fact that there are no clearly defined temples, or houses that are significantly larger than the others, argues for a fairly even distribution of wealth, according to Ian Hodder of Cambridge University, who is currently the lead archaeologist at the site. Rather, the people of Çatalhöyük seem to have retained many aspects of their egalitarian hunter-gatherer past. But its size—which is thought to have been in large part due to wealth gained through the trade in obsidian, found in large quantities in nearby mountains—meant that such an egalitarian social structure couldn't last.

The basic economic unit at all early Neolithic sites like Çatalhöyük was the nuclear family. It seems to have been how people apportioned farmland, stored their grain, and even practiced their religion. At Çatalhöyük, for instance, there seems to have been a well-developed ancestor cult in which people would bury the bones of their departed family members in the house, right under the floor. It was a way of clearly staking a claim to the location, and it reflected their sedentary and family-focused culture.

The formalization of religion during the Neolithic period likely proceeded slowly and took shape with the increasing stratification of society. Although we have no way of knowing for sure, it's likely that these early religions were based on formalizations of beliefs that had existed prior to the development of agriculture. The world's hunter-gatherers are traditionally animists, and their belief in a multitude of spirits and gods mirrors their reliance on a complex variety of natural resources. Agriculturalists, with their relatively simple food supply and their view of nature as something that needs to be *controlled* rather than *cooperated* with, were sociologically predisposed to create religions with fewer, more powerful gods—and gods in their own image at that. The first widespread images of humans appear only in the latter part of the Upper Paleolithic, with the near-universal presence of Venus figures in the archaeological record of Europe at this time. Such figures continued to be important during the Neolithic period, perhaps because humans recognized their increasing power. Their importance may have risen in concert with the development of the so-called goddess cult of the era, "goddess" being the name given to the female figures placed prominently at sites like Çatalhöyük.

These Venus figures, clearly symbolic, with their exaggerated breasts and hips, may have reflected the high social standing of women as givers of life. In a prescientific society, it's possible that the link between sexual activity and reproduction may not have been understood; if so, the birth of a child must have seemed supernatural. Furthermore, if women were the first to cultivate plants, their prominent role in coaxing life from the soil would have been clearly recognized. All fertility, whether from the loins or the land, sprang from them. Does this mean that women had higher social status than men—that is, were these early Neolithic societies matrilineal, with inheritance coming down the female line rather than the male? Perhaps, though some archaeologists (Ian Hodder, in particular) have argued against this interpretation, suggesting that

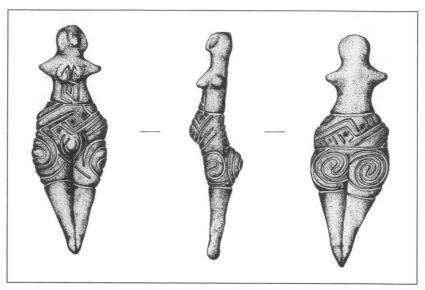

FIGURE 12: TERRA-COTTA VENUS FIGURINE FROM BULGARIA, CIRCA 5000 B.C. SOURCE: MARIJA GIMBUTAS, *THE LIVING GODDESSES*, UNIVERSITY OF CALIFORNIA PRESS, 1999.

men and women—at least in Çatalhöyük—were of similar status. Clearly, however, this didn't last.

As humans expanded their population size, they were forced to move from the original centers of domestication, mountain valleys, out onto the plains, since the small land area near easily accessible water supplies could not sustain an unlimited number of people. This necessitated the development of a system for transporting water and irrigating fields. Constructing irrigation canals requires that large groups of people work together toward creating a common, shared piece of real estate. This meant they had to develop some way of administering the work itself, as well as the maintenance of the completed canals and the access to the water—suddenly a scarce resource. And this meant they needed something else that had never existed before in human history: a formal government, with specialized bureaucrats and, most important, authority. Otherwise, why listen to someone telling you what to do?

Because of a growing population that was increasingly urban, as well as the growth in trade between early Neolithic cities, the need to oversee complex public works projects drove the development of ever more complex forms of government. The resulting growth of material wealth—surpluses of food and expansion in the amount of real estate that could be controlled by one family—showed that some members of society were inevitably better than others at tasks that mattered to the society. In other words, if you had a particular set of skills that were highly valued, you had a higher social status, which increasingly came to be reflected by the material rewards you accumulated.

The shift to wealthy urban settlements with goods to trade made the population more vulnerable to attacks from outsiders. After all, if you see a neighboring city becoming richer than your own, what's to stop you from attacking it and simply taking what its people have toiled so hard to accumulate? This would have led to the development of a formal military, rather than a citizen militia. There was suddenly something worth dying for. Because of its importance in defending the city from attack—and perhaps also, in aggressive cultures, its ability to go out and acquire new resources by attacking other settlements—the military would have played an important role in the society. And as with all skills, prowess in battle and military acumen would have been unevenly distributed. Since men are physically more suited to the intense physical activity of battle, they would have gained higher social status. The psychologist Simon Baron-Cohen has argued that men are naturally inclined to be better mechanical inventors than women (just as women, with their arguably better abilities to read subtle social cues, are probably better natural traders, at least judging from the preponderance of women in markets in the developing world), and technology was crucial in the Neolithic arms race. Either way, the new, warlike environment that came into existence during the Neolithic Revolution created an environment that was tailor-made for men to usurp the power of the goddess cults, or at least the equal status of women.

Although performing a seemingly trivial act, the first person to plant a seed set all of this in motion by tying our fortunes to those of the planted fields. Food became a fuel—a sort of primitive biodiesel, if you will—for powering social change. This was our first encounter with something I'll explore more fully later in this book—something I call *transgenerational power*. It is the idea that, with the increase in our power over the world around us brought about through the development of agriculture, we gained the power to affect events many generations down the line. Our minds, having evolved over millions of years during which we were hunter-gatherers, our only concerns about cause and effect extending perhaps a season into the future (will moving there make it harder to find game or gather plants in the dry season?), were not equipped to imagine the sequence of events that might occur long after we were gone. And given that as hunter-gatherers we were very much a part of the environment in which we lived, rather than in control of it, we didn't *need* to worry about how things might play out many years down the road. Our actions, like those of any other animal in the ecosystem we lived in, didn't have a large enough effect to perturb the "balance of nature." The development of agriculture changed all of this.

The events of history, now that we have entered the realm of the written record, are well chronicled, and there is no need to reiterate them in detail here. The first writing developed in Mesopotamia for recording transactions between traders, but it soon allowed people a measure of immortality by giving them a way to note details about themselves that would live on for posterity. The Palette of Narmer, an inscribed stone tablet dating from 3200 B.C., is the earliest known example of Egyptian hieroglyphic writing. Recording the unification of Upper and Lower Egypt by the pharaoh Narmer, it is perhaps the earliest description of a real-life historical figure. It is displayed prominently in the Egyptian Museum in Cairo, and every year two and a half million people have a chance to read about this person who lived over 5,000 years ago. Narmer must have been hoping for this kind of temporal influence when he commissioned the tablet, and it is a clear sign of how far humans had come in the few millennia since agriculture was developed. Villages had coalesced into cities, which had been joined together into empires with written records of their deeds to pass down to future generations. What before may have been lost to posterity or decayed into vague myth was now written in stone, an immutable statement about the person and the society he lived in.

Between 3000 B.C. and the Middle Ages, human society continued to grow and evolve along this same path, with ever more complex technology—bronze, then iron; mounted cavalry; powerful compound bows that greatly extended an archer's range—and larger empires: Assyria, Persia, Greece and Rome to the west, the Han and Khmer Empires in the East, the Mauryans in India, Great Zimbabwe in Africa. Religion became more formalized, with monotheism replacing polytheism among the majority of the world's population. Two of the world's great monotheistic religions, Christianity and Islam, with their great zeal for conversion, spread over wide territories to dominate the world. Their adherents soon numbered in the tens of millions—a far cry from the localized ancestor cults of the early Neolithic. But all of these events unfolded with a certain predictability, a natural outcome of what had been set in motion by the Neolithic Revolution.

Then, rather suddenly, another revolution happened. It had its roots in the growing importance of trade in the fortunes of states, which was replacing agriculture, religion, and rigid social stratification as the dominant sources of power. Whereas in the past people had often believed that they were preordained to live a particular type of life by God or king, by the eighteenth century it was becoming clear that this system was not necessarily producing an optimal outcome for the majority of the population. The Enlightenment works of Rousseau, Voltaire, Hume, and others questioned the status quo in a way no one ever had before, asking deep philosophical questions about the innate rights of human beings to life, liberty, and property. In Britain, in particular, the merchant class chafed at the limitations of life in a class-based society that still valued land over industry, despite the ever-increasing importance of the latter.

The speed at which the Industrial Revolution played out was a testimony to how interconnected the world had become by the nineteenth century. While it had taken nearly 10,000 years, ending with the European colonial era, to bring the seeds of agriculture to every

last corner of the globe, it would take only a few generations to dis-
tribute the factories of the Industrial Revolution. This is because the
rate at which information was conveyed had increased exponentially
since the days of Abu Hureyra and Çatalhöyük. It is possible that at
the dawn of the Neolithic, so localized were the lives of each village
that each would have spoken its own language or dialect. The consol-
idation of empires in the Bronze and Iron Ages would have enforced
a dominant language on the groups that were conquered, thus facili-
tating the spread of information (and, therefore, making the empire
governable). This common language would also have led to a much
faster spread of innovations from one place to another, which would
have increased the rate at which innovation occurred. By the middle
of the nineteenth century, the majority of the world's population
spoke, at least as a second language, one of three European languages:
English, French, or Spanish. This was the environment that allowed
the Industrial Revolution and its ideas to spread like wildfire.

Inevitably, the Industrial Revolution and its reliance on the
rapid transmission of ideas led to the modern era, where access to
information is the key to power and wealth. Of the world's population of 6.8 billion, it is estimated that 25 percent, or 1.7 billion
people, use the Internet; this is more than the total number of people on the planet in 1900. In North America, nearly three-quarters
of the population uses it regularly. It's not surprising, then, that we
are currently experiencing a period of innovation that makes even
the Industrial Revolution pale in comparison. Ideas fly from
Boston to Bangalore to Brisbane in milliseconds, carried through
pipes that drip only o's and 1's from their spigots. It should be the
best of times, with ever-increasing life spans and ever-richer people. But it isn't. As the quote by Richard Tomkins that opens this
book suggests, life today is far from perfect, and many people feel
that things are getting worse. In the next chapters I'll examine
some of the problems we have come to face during this ten-
millennium period of revolutionary change, and will show how
they ultimately trace back to a mismatch between the culture we
set in motion 10,000 years ago by planting that first seed and ourselves as human beings.

# Chapter Three
# Diseased

*Infectious disease which antedated the emergence of humankind will last as long as humanity itself, and will surely remain, as it has been hitherto, one of the fundamental parameters and determinants of human history.*

—WILLIAM McNEILL,
*Plagues and Peoples*

*Overeating and inactivity are the proximal causes of obesity, but unfettered consumerism drives the obesity pandemic of the twenty-first century.*

—ELLEN RUPPEL SHELL,
*The Hungry Gene*

## DOLLYWOOD

The Great Smoky Mountains sit toward the southern end of the Appalachian Range, which runs nearly the entire length of the eastern coast of the United States, from Labrador to northern Alabama. Thrust upward around 450 million years ago by the collision of continental plates during the creation of the supercontinent of Pangaea, the Appalachians were at one time part of the same range as the Atlas Mountains in North Africa. With the highest peaks in the entire chain, the Smokies have long been a magnet for outdoor enthusiasts from the surrounding states. Hiking through the lush spruce forests interspersed with native rhododendrons, or skiing down the only slopes in the southeastern United States with reliable natural snow, the large number of visitors make these mountains a hub of activity.

In the past fifty years the towns of Gatlinburg and nearby Pigeon Forge, the main settlements on the Tennessee side of the park, have benefited from an ever-increasing flow of tourists, which topped 9.4 million in 2006, making Great Smoky Mountains National Park the most visited in the country. Pigeon Forge and

Gatlinburg have recently eclipsed the park, though, and attracted more than 11 million tourists that same year. While Gatlinburg has to a certain extent maintained its 1950s small-town feel, Pigeon Forge has seen nearly unchecked growth. It is now a sprawling, traffic-clogged tourist destination, one of the largest in the southern United States. Although it has outlet malls, go-kart racing, and other typical "destination" attractions, its number one tourist draw is Dollywood, an amusement park co-owned by Dolly Parton, the buxom country-and-western singer, which attracts over 2 million visitors every year.

I visited Dollywood on a hot, sunny day in late June, near the peak of the summer tourist season. In between taking my children on roller coasters and waiting in line for ice cream, I had a chance to observe the surroundings—the human surroundings, to be precise. What struck me most was the noticeably high level of obesity; many people were not just overweight but clinically obese—that is, they had a body mass index (BMI) of 30 or greater, which is equivalent to being five foot nine and weighing more than 205 pounds. Having lived on the East and West Coasts of the United States and in Europe for the past two decades, I wasn't used to seeing so many large people. The rides with fitted seats and passenger-retention devices had "sample seats" near their entrances so people could see if they would fit into the ride—and many didn't. While the corn dogs and funnel cakes certainly didn't help keep waistlines svelte, this plethora of paunch was clearly part of a deeper and more long-standing problem.

Although people come from all over the country to visit Dollywood, it primarily draws visitors from nearby states—Tennessee, Kentucky, Virginia, West Virginia, Alabama, Mississippi, and other parts of the South and the Midwest. According to recent data on the patterns of obesity in the United States published by the Trust for America's Health (TFAH), these are among the fattest states in the country. In Mississippi, for instance, nearly a third of the adult population is clinically obese, making it the heaviest state in America, and West Virginia and Louisiana are close behind. When the fifteen fattest states are mapped, a striking pattern emerges.

**FIGURE 13: THE FIFTEEN U.S. STATES WITH THE HIGHEST LEVELS OF OBESITY.**

The clustering of obesity in the South and the Midwest is striking, and not at all what would be expected to occur by chance. There are certainly genetic factors involved in obesity, but genetics fails to explain both this geographic clustering and the increasing rate of obesity over the past two decades in 1991, no state exceeded an obesity rate of 20 percent. Our genes haven't changed that much in less than a generation. Clearly, another explanation is needed.

The answer likely lies in the relative socioeconomic status of the people living in these states. Mississippi, Arkansas, and West Virginia are the poorest states in the country, with per capita incomes around half of that in the richest state, Connecticut. South Carolina, Alabama, Louisiana, and Kentucky are also in the bottom eleven states. The poorest states also trail in the rankings of educational attainment, with around 20 percent of the population not completing high school. This is certainly part of a vicious cycle: high school dropouts tend to have significantly lower incomes than people who complete high school or college, and poorer children are less likely to complete high school. But why does being poor

and uneducated put you at higher risk for becoming obese? It
seems that, if anything, poorer people should be *less* able to afford
enough food to make them fat.

By now probably everyone in the world has heard about the
problem of obesity, so it shouldn't come as news that there are a lot
of fat people out there. Films and books like *Super Size Me* have
brought this problem to the public in a very visible way, and according to the same TFAH survey, 85 percent of Americans now
consider obesity to be an epidemic. According to the United States
Centers for Disease Control and Prevention and the World Health
Organization, obesity is now the second most important root cause
of illness and mortality, after cigarette smoking, and may overtake
smoking in the next ten years, as more and more people around the
world become overweight. Our food, meant to nourish us, is now
killing us.

If obesity is an epidemic, though, it is clearly affecting some people more than others. The international trend mirrors the one in the
United States, with richer countries tending to have lower obesity
rates. In Europe, for instance, there is a weak but noticeable trend
for countries with lower average levels of education and per capita

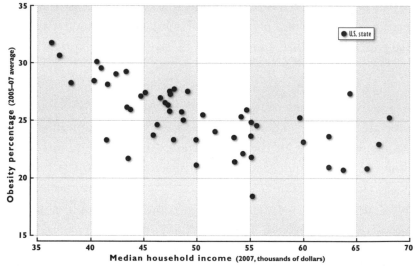

FIGURE 14: THE INVERSE RELATIONSHIP BETWEEN OBESITY AND HOUSEHOLD INCOME IN THE UNITED STATES.

gross domestic product to have higher levels of obesity; levels are typically higher in countries of the former Eastern bloc and those in the eastern Mediterranean, such as Greece and Cyprus. The pattern is not as clear as it is in the United States, but it is suggestive.

Several reasons have been posited for this odd correlation, ranging from lack of understanding of the relationship between obesity and disease among the less educated to lack of access to healthful food. It is true that wealthier people in the developed world are more likely to eat fresh, nutritious foods, and in the past few years there has been a strong movement in some upper-middle-class circles for a return to locally grown, organic ingredients. Many people are not aware of, or ignore, the dangers associated with the increasing proportion of highly processed, sugar-laden "fast food" so readily available in the developed world, as Eric Schlosser has pointed out in his fascinating and frightening book *Fast Food Nation.* Couple this with a lack of physical activity as more and more people leave manual jobs in factories and farms for service-sector jobs with long commuting times, and it is a recipe for disaster.

While the poorest in the developing world remain relatively thin due to actual food shortages and the high levels of everyday physical activity necessary for mere survival, as they become richer, some members of even the world's poorest countries fall prey to the adiposity trap. It seems that the key to producing an obesity epidemic is for a society to have just enough money and education to give it access to an excess of processed food and an inactive lifestyle, but not enough for it to realize that overeating and not exercising is dangerous. This "danger zone" of development, as it spreads into the developing world, is expected to increase the global burden of obesity enormously in the coming decades. As more people in their countries enter the ranks of the global middle class, Indian and Chinese government officials have become increasingly concerned about their rising obesity rates, which have doubled in the past decade, largely in the wealthier urban populations.

The world's policy makers and public health experts recognize the problem, but dealing with it is another matter. Diabetes, one of the serious chronic diseases stemming from obesity, is now a common fact of life in places where a generation ago people were still starving. It is "the price you pay for progress," according to the director of a hospital in Chennai, India, who specializes in the ailment. His resignation stems from the difficulty of getting people to change their lifestyles. Fattening food and inactivity are powerful drugs, it seems, and appeal to something deep in our nature, as though we were genetically programmed to fall under their spell. Which, in many ways, we are.

## THRIFTY GENOTYPES AND PROFLIGATE TASTES

James Neel was an American physician and geneticist who spent much of his career working with the Yanomami Indians of the Brazilian Amazon. He had started out studying evolution in fruit flies, but he decided to go to medical school soon after joining the Dartmouth faculty, in 1939, and ended up specializing in human genetics. After spending the years after the Second World War studying the genetic effects of the Hiroshima and Nagasaki atomic bombs (surprisingly, he found that there was no increase in the mutation rate among those exposed to the radiation from the bombs), he decided to combine his early interests by applying genetic methods to the field of anthropology.

The Yanomami fascinated Neel because they had only recently been contacted by the outside world, meaning that they were presumably living in a state of nature—in effect, the way our ancestors probably lived. Neel thought that by studying such populations scientists could gain an insight into the past selective pressures on the human genome, which (as we have seen in Chapter 1) changed significantly during the transition to agricultural life. One of the main environmental changes, Neel thought, had to do with the increase in readily available food in agricultural societies. Although we now know that Paleolithic hunter-gatherers rarely led lives of privation, and may have had better access to a wide variety of food than their agriculturalist neighbors, it is true that the diet of agriculturalists is fundamentally different from that of the typical hunter-gatherer. In general, agriculturalists have a much higher percentage of carbohydrates—sugars and starches—in their diet. In China, for instance, the large quantities of rice consumed yield a diet that is about 75 percent carbohydrate. High levels of physical activity, however, mean that obesity is less of a problem in largely rural China. In Chinese cities, though, obesity is on the increase, as is diabetes.

Neel suggested that diabetes, rare among hunter-gatherers, is a physiological reaction to the sudden increase in easily available calories. What might have been highly adaptive to a hunter-gatherer—the ability to maintain physiological function under conditions of low caloric intake—could be maladaptive with a richer diet. He called this the "thrifty genotype," and it has become a widely accepted hypothesis for the widespread existence of diabetes, which should otherwise, due to the action of natural selection, be extremely rare.

Diabetes comes in two forms. Type 1 usually manifests itself in childhood; it is the one that is treated using insulin shots. It is caused by a complex interaction between inherited susceptibility factors—DNA variants—and the environment in which the child is raised. Type 2 is more complicated and usually occurs in adulthood (although teenagers and even children are now showing up with this form as well). It is caused in part by genetic factors but also by significant environmental effects, particularly diet. In particular, more than 80 percent of people with type 2 diabetes are overweight.

The best example of the negative effects of the thrifty genotype is found among Pacific Islanders and Native Americans. Samoans, for instance, settled their islands over 3,000 years ago, during the so-called Polynesian expansion from Southeast Asia into the Pacific. The ocean voyages involved in these extraordinary migrations would have placed people under intense physiological stress for weeks at a time, and it is possible that there was strong selection for people who were able to reduce the rate at which they burned calories. This thrifty genotype would have been a fantastic thing to have on the voyage, but it would have been more of a problem once the group arrived at their destination. However, as with other subsistence agriculturalists around the world, the high levels of physical activity involved in everyday life until quite recently would have prevented an obesity epidemic.

With the arrival of modern civilization, however, the Samoans have stopped tending fields and fishing from paddled outrigger canoes, and now spend much of their time inactive. Particularly in American Samoa, their diet, once rich in fish and vegetables, has become more like that of the average mainland American, and is primarily composed of imported foods like Spam (which entered the Samoan diet via the United States military during the Second World War) and other highly processed delicacies. The result is that nearly two-thirds of urban Samoans are now clinically obese, and even their rural counterparts have obesity rates of around 50 percent. This might produce fantastic sumo wrestlers (one of the top wrestlers in the 1990s, Musashimaru Koyo, was Samoan), but it also gives this tiny island nation one of the highest incidences of diabetes in the world—25 percent in men and 15 percent in women. Sometimes being thrifty can kill you.

Similarly, the Pima Indians of the southwestern United States and northern Mexico show a frighteningly strong correlation between lifestyle and diabetes. Known for having the highest incidence of diabetes in the world—around 40 percent—the Pima living in the United States have, like the American Samoans, succumbed to the lure of a modern lifestyle. Their cousins across the border in Mexico, however, with a much more traditional way of life and higher levels of physical activity, have a correspondingly lower (though still relatively high) incidence of diabetes, around 7 percent. Clearly, something about the American lifestyle is more than quintupling the Pima Indians' diabetes risk. That something has a lot to do with the large number of calories in the typical American diet, many of them in the form of processed carbohydrates and fats, as well as the lower levels of activity in our sedentary culture.

Overeating and low levels of physical activity, leading to obesity, and cigarette smoking, leading to hypertension and cancer, are the major causes of preventable death in the world today. As my own doctor says, "It's really not that complicated—exercise and don't smoke and you'll be healthy." But such a prescription is not so easy to follow, as we have a couple of things working against us in the form of evolutionary baggage. First, the desire to eat is a basic survival instinct, so it's unnatural to try to reduce the amount of food we consume. Second, our hunter-gatherer ancestors would have found the idea of exercise for exercise's sake ludicrous—their goal, as with most hunter-gatherers today, was to minimize the amount of energy they expended when going about their everyday activities, because every calorie they wasted had to be replaced by finding more food. Finally, chemical compounds like nicotine actually mimic substances that occur naturally in our bodies, which is why our nerve cells have receptors on their surfaces that are stimulated by them. It's just bad luck that a molecule that evolved to protect the tobacco plant from pests turned out to be a stimulant to humans, and that the easiest way to get this compound into our systems—burning the leaves of the plant—also produces high levels of tar, which is the primary culprit behind smoking's cancerous effects.

What clearly seems to be happening is a profound shift in the causes of disease from threats from without to threats from within. More and more, we are causing our own deaths, rather than succumbing to some other force that is largely beyond our control. This can be seen in the increased focus of public health agencies on chronic diseases like diabetes. America's Centers for Disease Control and Prevention, for instance, was originally founded in 1942 to deal with malaria, then a widespread threat in the southern United States. Later it focused on vaccinations and emerging infectious diseases, but a significant amount of its work these days is on the far greater threats of noncommunicable diseases. Some of this work is epidemiological, but much of it is spent communicating the risks of poor diet, lack of physical activity, and smoking. The informational pamphlet and, increasingly, Web-based communications have become the twenty-first-century equivalents of the smallpox vaccine. Unlike a vaccine, though, protection is granted only through the active participation of the patient population over their lifetimes—something that is clearly less likely than a predictable immune response to a vaccination.

A recent report compiled by the CDC and the nonprofit institute RTI International blames obesity for much of the massive increase in health-care costs in the United States over the past decade. Overall, they calculate that obese people average $4,871 in medical bills each year, while people with a healthy weight cost $3,442. According to health economist Eric Finkelstein, the lead researcher on the report, "obesity is the single biggest reason for the increase in health care costs." The increase in American obesity is not only a public health hazard—it's an economic time bomb as well.

At Dollywood, I got a view of the future—a possible future, where the human form has changed to such an extent that people waddle rather than walk and can't sit in seats designed to accommodate even fairly large frames, where even minor physical activity produces breathlessness, and where the ingredients in the food sound more like a chemistry lesson than something you'd consider eating. But solving the obesity epidemic, with its complex interactions between genetics, income, education, and culture, requires more than simply prescribing a daily walk and fewer cheeseburgers and sugary sodas. It requires an understanding of the long-term history of human disease, and how friends—the plants we cultivate—ultimately became enemies.

### THREE WAVES

In late February 2003, a Chinese doctor traveled to Hong Kong to attend his nephew's wedding. He was sick at the time, but he put it down to a cold or perhaps the flu. He certainly didn't suspect that he would inadvertently produce a global epidemic that would ultimately infect thousands of people around the world, lead to the deaths of hundreds, and cause a noticeable drop in world economic output, equivalent to around $10 billion. Tourism revenue—plane tickets, hotel stays, meals out—dropped by 9 percent worldwide, and to this day visitors to Singapore and some other Southeast Asian cities are examined by an infrared temperature scanner as they pass through security, like something out of a science fiction film. Too red a face indicates a fever, which will lead to further questioning by the medical authorities to ascertain whether the traveler might be carrying a dangerously infectious disease.

The doctor had unknowingly been infected with a new virus, one now understood to be a member of the coronavirus family, whose members include viruses that cause the common cold. But our Chinese doctor's illness was no case of the sniffles. People infected with the new virus spiked high fevers, eventually developed pneumonia, and, in around 10 percent of the cases, died by drowning in their own lung fluid. Severe acute respiratory syndrome, or SARS, had been born.

The most amazing thing about SARS is the speed with which it spread. What probably began in late 2002 when a chicken or pig virus in southern China hopped from its animal host to a human—possibly a food worker—had spread by late March 2003 to countries as far away as Canada, Switzerland, and South Africa. All of the people infected with the virus initially had spent some time in Southeast Asia, but secondary infections became more common in April as the virus spread from the primary infectees to their friends, family, and hospital staff. This is because of the ease with which it was transmitted—essentially like a cold virus, through sneezing and contact with an infected person.

In the grand scheme of things, SARS was not a major killer. Far more devastating were the fears and paranoia it engendered, which contributed to the failure of many Asian businesses and several airlines in 2003. The death rate, although three to four times higher than that of the influenza virus that caused the 1918–19 global epidemic, in which around twenty million people died, was not as high as those of many other diseases. Among these are Ebola, whose outbreaks have been limited thus far to remote locations in central Africa, Lassa fever, Marburg virus, and others with death rates approaching 100 percent—a far more sobering figure, although the isolated nature of their outbreaks and their difficulty of transmission makes them somewhat less threatening.

Not so for the much-discussed potential pandemic threat of so-called H5N1 avian flu, which has resulted in the death of more than half of the people who have been infected so far. Avian flu, as the name suggests, was originally transmitted to humans through close contact with birds. This type of contact occurs most often on farms and in markets, and its initial emergence in China and Southeast Asia is thought to reflect the especially close contact in this region between humans and their domesticated ducks and chickens, something not as common in many other parts of the world today, where factory farming has become more widespread. Similarly, the 2009 spread of a much less deadly form of swine flu—similar in its structure to the 1918 flu strain—has led some experts to suggest that we are on a collision course with a new plague of biblical proportions.

It turns out, interestingly, that the "new threat" of deadly diseases spread through close contact with animals is not new at all. In his influential 1976 book *Plagues and Peoples,* Canadian-American historian William McNeill examined the impact of disease—particularly epidemic disease—on human history. Disease, McNeill argued, has long been a catalyst for significant historical events, and he summoned groundbreaking evidence to explain the role of the Mongol Empire in Europe's fourteenth-century black plague epidemic and the importance of Eurasian diseases in allowing Spanish conquistadors to subdue the empires of the Americas. Jared Diamond's *Guns, Germs, and Steel* is perhaps the best known of the many later books influenced by McNeill's work.

In *Plagues and Peoples,* McNeill traces the origin of many diseases common today back to changes in human society during the Neolithic period. Many of these changes we are familiar with from the last chapter, including the increasing number of people living in a relatively small space, allowing rapid transmission of diseases by infected individuals, and a large enough pool of uninfected people to permit the emergence of epidemics. Perhaps the most important factor, though, was the domestication of animals. As the human population increased in early farming communities, hunting was no longer a viable option—as with wild seed-bearing grasses, the supply of wild animals was limited by the natural carrying capacity of the land. This meant that many were soon hunted to near extinction. The necessity of creating a stable food supply led human populations in the Middle East to begin domesticating sheep, goats, pigs, and cattle from their wild progenitors by around 8000 B.C., and the Southeast Asian population to domesticate the chicken by around 6000 B.C. This created a reliable source of meat in the Neolithic diet, but the large numbers of people and animals cohabiting also created an environment that had never before existed in human history.

For the first time, people and animals were living in the same communities. While Paleolithic hunters had certainly come into contact with their prey after a successful hunt, the number of wild animals contacted was a small fraction of those living in the newly domesticated Neolithic herds. Also, most of these animals were dead; this would have decreased the chances of transmitting many diseases, but perhaps facilitated the transfer of blood-borne infections. When we started living close to animals throughout our lives—particularly as children—the odds of diseases being transmitted increased significantly. Although some such infections had probably always existed to a lesser extent in both the animal and human populations, suddenly there was a brand-new opportunity to swap hosts. The microorganisms had a field day.

McNeill wrote that of the diseases shared by humans and their domesticated animals, twenty-six are found in chickens, forty-two in pigs, forty-six in sheep and goats, and fifty in cattle. Most of the worst scourges of human health until the advent of vaccination in the eighteenth century were imports from our farm animals, including measles, tuberculosis, smallpox, and influenza. Bubonic plague was transmitted to us by fleas from rats living in human settlements. As far as we can tell from the archaeological record, none of these so-called zoonotic diseases (from the Greek *zoon,* for animal, and *nosos,* for disease) afflicted our Paleolithic ancestors—all seem to have arisen in the Neolithic with the spread of farming. McNeill suggests that many of the plagues described in the Bible may coincide with the explosion of zoonotic diseases during the emergence of the urban civilizations of the Neolithic, Bronze, and Iron Ages.

What is clear is that a new source of human mortality had arrived on the scene. This does, however, raise the question of what people had been dying of before the development of agriculture. Were there really no diseases in the human population? Of course there were. It's likely that macroparasites—things such as tapeworms that can be seen by the naked eye—were problems for our distant ancestors. Most of these infections generally would have produced little beyond feelings of malaise, though—not acute, debilitating symptoms like high fevers, organ failure, and death—in part because we had probably been evolving together with these parasites for such a long time. Over millions of years, an evolutionary process known as mutualism would have led the parasites to produce less acute physical symptoms in their hosts (us), since it does a parasite little good to kill its host and thus its source of food, and we would have adapted to their presence. In general, the longer an infection has been around, the less virulent it is, the symptoms it elicits in the host becoming less severe over many generations. New diseases that erupt suddenly into a previously unexposed population often have extreme outcomes, including death.

If macroparasites couldn't have produced a significant amount of mortality during the Paleolithic period, and most disease-causing microorganisms hadn't yet had a chance to pass from animals to humans, what did our hunter-gatherer ancestors die of? According to British evolutionary biologist J.B.S. Haldane, traumatic injuries were the most likely cause of death throughout most of human history. Does this mean we spring from a race of klutzes, who tripped and fell their way through the Paleolithic? No: such injuries would have included wounds sustained during hunting and skirmishes with other groups, the traumas associated with childbirth (a significant source of mortality for both mother and child until quite recently), and accidental falls and drownings. All of these hazards, coupled with infections from the wounds, would have been the main cause of hunter-gatherer morbidity and mortality.

So, we seem to have evidence for an interesting pattern—three waves of mortality as we move from Paleolithic times to the present. The first is trauma, primary from the time of our hominid ancestors until the dawn of the Neolithic period. As people settled down and began to domesticate their animals rather than hunt them, infectious disease began to supersede trauma as a significant cause of mortality. This second wave, of infectious disease, continued to be the most significant cause of death until antibiotics were developed in the mid-twentieth century. The final wave has happened since the mid-twentieth century, in developed countries, where vaccinations and widespread antibiotic use have reduced infectious diseases to a fraction of their former threat. Now that we

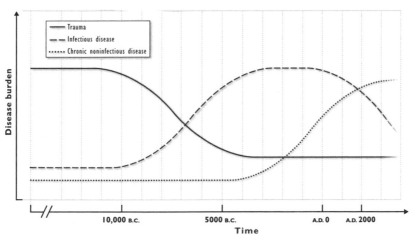

FIGURE 15: THE CHANGING "WAVES" OF DISEASE BURDEN OVER THE PAST 15,000 YEARS.

have stemmed the joint threats of trauma and infection, chronic diseases are becoming a larger threat. Most people prior to the twentieth century would have died relatively young, before these maladies—particularly diabetes, hypertension, stroke, and cancer—would have had a chance to develop. With modern medicine we've traded the scourges of trauma and infection for a threat from within our own bodies.

This progression from one disease threat to another seems to be one of the laws of human development, and lets us make predictions about what will happen as the world's less developed countries move from subsistence agriculture, with a way of life not unlike that of our ancestors during the Neolithic period, to enjoying the wonders of the modern world. At the moment, however, infectious disease is still the main threat to most of what used to be called the Third World. And one infectious disease in particular kills more people than any other in the world, more than two million per year, 90 percent of whom are African children under the age of five: malaria. But malaria, unlike most other infectious diseases afflicting us, did not originate in one of our domestic animals. It was probably around, albeit at lower levels, even in the Paleolithic. Its rise to infamy illustrates another continuing effect of the Neolithic Revolution, and shows how human architectural and city planning decisions can have disastrous unintended consequences. The evidence comes not from the archaeological record, though, but from DNA.

## LANDSCAPING THE GENOME

It is perhaps fitting that the man who is credited with discovering Angkor, the magnificent temple complex in Cambodia, died of malaria in the jungles of Laos in 1861. Henri Mouhot, a French explorer, had spent the previous three years exploring the monuments of Thailand, Cambodia, and Laos, and certainly popularized them for Western readers in his book *Travels in the Central Parts of Indo-China, Cambodia and Laos During the Years 1858, 1859 and*

*1860,* published in 1863. Of course the local people knew about
the existence of the monuments, but in the age of colonialism
Mouhot was credited with being the first to discover them. His po-
etic descriptions led to restoration efforts, and to this day they remain one of the world's great sights.

The Angkor complex was built in stages by the rulers of the
Khmer Empire, the most powerful in mainland Southeast Asia, between the ninth and fifteenth centuries. It includes the famous
temples of Angkor Wat and Bayon, with their huge stone con-
structions and immense Buddha heads. Satellite surveys have recently revealed the extent of Angkor to be a staggering 380 square
miles, the largest preindustrial city in the world—larger even than
modern New York. This complex of temples and houses was home
to more than 750,000 people at its peak, but was abandoned in the
fifteenth century, all of its temples left to the creeping jungle except for Angkor Wat.

Several theories have been advanced for why its residents de-
serted this huge city, but no one knows for sure. A war with the
Thai Empire, based northwest of the Khmer, is thought to have

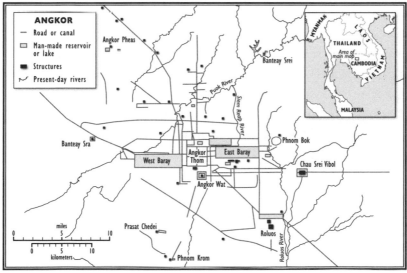

**FIGURE 16: ANGKOR, THE LARGEST PREINDUSTRIAL CITY IN THE WORLD.**

contributed, but most scholars now think that ecological stresses had a deciding effect on the fortunes of the Khmer capital. To feed 750,000 people and support the complex political and religious infrastructure of Angkor required enormous quantities of natural resources, as with any city today. The most important of these would have been water, which was necessary not only for drinking but also for the wet-field rice agriculture that fed the city's population. To provide such enormous amounts of water, a complex system of canals carried water from the Puok, Roluos, and Siem Reap Rivers to huge reservoirs known as *barays*. And to grow the rice, large portions of Angkor were given over to open fields.

It seems that climatic shifts in the Northern Hemisphere associated with the end of the medieval Little Ice Age period in the fourteenth through seventeenth centuries may have changed the monsoonal pattern in Southeast Asia enough to have led to a shortage of water for Angkor, which would have led to some of the rice paddies being abandoned when their water usage became untenable. Religious changes during this period, perhaps in part a response to the more challenging climatological conditions, also probably led some residents to leave the Khmer capital. But the human-engineered landscape probably contributed in an unforeseen way.

Malaria is the best-known vector-borne human disease ("vector-borne" means that it is transmitted from one person to another by another animal—the "vector"). In this case, the vector is a mosquito, in particular those belonging to the genus *Anopheles*. Female *Anopheles* mosquitoes, in order to have enough amino acids to produce eggs, drink blood from mammals through modified mouth parts that function as a kind of hypodermic syringe. In order to keep the victim's blood from forming a clot that would clog the insect's blood-collection system, the mosquito injects a small amount of her saliva, containing anticoagulants, into the tissue around the bite. This allows her unfettered access to the animal's blood, and it also serves as a wonderful opportunity for microorganisms to move from host to host.

We now know that malaria is caused by protozoa of the genus

*Plasmodium.* These small multicellular organisms were probably originally spread, like most other infectious diseases, through direct ingestion of infectious material, such as waste or tainted water. However, at one point one of their distant ancestors developed the ability to infect red blood cells, which opened up a whole new route of transmission. Infected animals, when bitten by a bloodthirsty *Anopheles* female, would have their parasite-tainted blood transmitted into the mosquito's salivary glands, which would then be injected into the next animal bitten. A wonderful example of evolution in action, and the root cause of the scourge that now kills millions of people every year.

Until the twentieth century, though, malaria was thought to be caused by a bad landscape—*mala aria* is Italian for "bad air," particularly the damp air around swampy areas, and it was this that early European explorers sought to avoid. Mosquitoes were certainly annoying, but their link to the awful fevers brought on by this killer disease was not recognized until the era of modern medicine. However, in a way the early Europeans were correct: bad geography did lead to the disease, although not through the poor air quality surrounding swamps. Rather, it was the presence of wet, dank mosquito breeding grounds in the swampy areas that made for a large number of hungry female mosquitoes, and therefore more likely malarial transmission.

In an article published in 1992, the French epidemiologist Jacques Verdrager suggested that as Angkor's rice paddies were abandoned, they served as a perfect breeding habitat for malarial mosquitoes, particularly the common Southeast Asian species *A. dirus.* As the malaria-transmitting species increased in numbers, the human population became widely infected, and a vicious cycle ensued. More people died or left Angkor, which led to more paddies being abandoned, and within a few generations Angkor had been nearly completely depopulated. While the degree to which this process played a significant role in Angkor's demise is uncertain, it seems plausible that a nagging insect helped to bring about the ruin of the greatest city in the world.

*Anopheles* mosquitoes are found worldwide, and nearly forty species can harbor *Plasmodium.* Largely tropical or subtropical species, they are particularly common in Africa, Southeast Asia, and Latin America. In part because of the mosquitoes' ubiquity and their role in transmitting malaria, many scholars of human prehistory have long assumed that malaria has been a scourge of humanity throughout most of its evolutionary history. Recent work in genetics, however, is revealing a more complex story.

Studies of DNA from *Plasmodium* parasites by National Institutes of Health researcher Deirdre Joy and her colleagues have revealed that worldwide populations of *Plasmodium falciparum,* which causes the most dangerous form of malaria, have been diverging for at least 50,000 years. This date suggests that early humans took African malaria with them on their journeys out of Africa to populate the globe, the earliest of which began at that time. What is perhaps more interesting is that Joy and her colleagues also found evidence of a massive expansion of *falciparum* malaria out of Africa within the past 10,000 years—the same time as the expansion of agriculture during the Neolithic period.

Another genetic study has revealed a complementary insight into a change in the recent past, but this time from patterns detected in the *human* genome. University of Pennsylvania geneticist Sarah Tishkoff and her colleagues, through a careful analysis of genetic variation associated with the *G6PD* gene, found evidence of strong selection acting on the gene in the past 10,000 years. *G6PD* is an enzyme that helps to convert glucose—sugars in the diet— into the subcellular energy packets known as nicotinamide adenine dinucleotide phosphate, or NADPH. NADPH is one of the energy currencies of the cell, and it and its biochemical brethren NADH and ATP are ultimately where all of the energy in your food ends up. *G6PD* is a rather important enzyme, in other words, and its function has been fine-tuned over hundreds of millions of years of evolutionary history. In some human populations, though, *G6PD* carries mutations that reduce its ability to function. Favism is the common name for the disease that results, so called because its symptoms often manifest themselves when fava beans are eaten.

These symptoms include anemia, jaundice, and kidney failure. Full-blown favism is a nasty disease, something you wouldn't wish on your enemies, let alone your children.

However, some reduction in *G6PD* function has an interesting side effect: because *G6PD* seems to be particularly active in red blood cells, the effects of a deficiency are felt most acutely there— exactly where the *Plasmodium* parasites are also active. It seems that during their reproductive cycle in the red blood cells, the malarial parasites siphon off the NADPH supply for their own uses—they are *parasites,* after all—and this stresses the cell's metabolism enough so that the cell essentially commits suicide, killing the parasites in the process. Children who inherit these defects in *G6PD* function, while inheriting the predilection to favism, also gain protection from the malaria parasite.

Tishkoff and her colleagues applied methods similar to those used by Jonathan Pritchard, which we learned about in Chapter 1, to look at variations surrounding the *G6PD* variants. They looked at two in particular, one common in Africa and the other in Mediterranean populations (typically found at frequencies of 20 percent or so, depending on the population). By assessing the pattern of genetic variation linked to the variants, the geneticists estimated the African version to be between 3,840 and 11,760 years old; the Mediterranean form appeared to be even younger— between 1,600 and 6,640 years old. In other words, both had arisen in the past 10,000 years. It was a startling discovery and, coupled with the results from the plasmodium genome, suggested that malaria had become a significant human scourge only within the past 10,000 years.

As those of you who have been paying attention will realize, this is precisely the time of the great changes in human society brought on during the Neolithic period. Malaria, a very old disease that probably afflicted hunter-gatherers living in the tropics tens of thousands of years ago, became a much greater threat once we settled down and started farming. While part of this increase probably stems from the increasing population density in farming communities, some of it also seems attributable to the farming methods themselves. In particular, the landscaping choices made by early farmers in malarial regions almost certainly led to a greater incidence of the disease. As at Angkor, the creation of open areas in the forest (in the case of tropical Africa) and reservoirs and slow-moving canals (in the Middle East) would have resulted in ideal breeding grounds for mosquitoes. The *Anopheles* mosquito needs shallow, sunlit water in which to breed, and there weren't nearly as many such pools in the preagricultural era. Once humans started to clear the forest and plant crops, these insects would have become much more common. By growing food and reengineering the landscape, it seems, early African farmers were probably sowing the seeds of a new epidemic as well.

Malaria, then, seems to fit the pattern of other infectious diseases, which increased in frequency during the Neolithic period. The second wave had arrived in full force, and its effects are seen today both in the genome of the disease-causing organisms and in our own genome. The increase in mobility over the past two centuries is stirring up the infectious disease pot at an alarming rate, and the airplane is today's equivalent of the irrigation canal. Emerging infectious diseases promise to be a serious challenge over the next few decades, as swine flu, new-variant Creutzfeldt-Jakob disease, and HIV all attest. But the longer-term battle will not be with these microorganisms; rather, it will be with our own biology. This third wave of chronic diseases is still cresting, and its rise from obscurity is where we're headed next.

## CARBS AND CAVITIES

The village of Mehrgarh in present-day Pakistan lies in the shadow of the Toba Kakar Range of the western Himalayas, whose peaks rise to over ten thousand feet. These mountains are extremely remote and rarely visited, and were thought in 2004 to be where Osama bin Laden was hiding from Western military troops. The Bolan Pass, the only relatively easy route through the range, has served as a gateway

to southern Asia for millennia. Travelers making the journey from
central Asia down to India would leave the mountains and find
themselves on the dry Kachi Plain to the west of the great Indus
River, which meanders two thousand miles from its source in Tibet
to the Arabian Sea. It seems an unlikely place to build a village, but
this is where Mehrgarh was founded—over 9,000 years ago.

Mehrgarh is one of the oldest Neolithic settlements in the
world, and the oldest in southern Asia. A precursor to the third-
millennium B.C. Indus Valley civilization, which included the extensive settlements of Harappa and Mohenjo Daro to the east,
Mehrgarh has been the subject of extensive archaeological excavations since the 1970s. As with Neolithic settlements in the Middle
East, the pattern is one of ever-greater reliance on domesticated
animals and plants over time. Judging from the oldest layers of
the excavations, people still hunted large game, but this ended
abruptly as domesticated animals made their appearance. The inhabitants of Mehrgarh raised wheat and barley, and they kept cat-

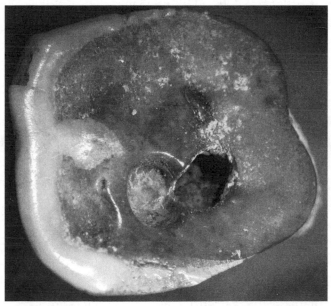

FIGURE 17: NEOLITHIC DRILLED MOLAR FROM MEHRGARH, PAKISTAN. (PHOTO COURTESY OF DR. LUCA BON-
DIOLI AND DR. ROBERTO MACCHIARELLI.)

tle, goats, and sheep. They lived in mud-brick houses and made pottery. They worked metal and traded with the surrounding regions—lapis lazuli from the Pamir Plateau, five hundred miles to the northeast, has been found, as have seashells from the Indian Ocean. The people of Neolithic Mehrgarh had what in many ways was a typical Neolithic lifestyle. They also had cavities.

One of the most intriguing things to come out of the work on Mehrgarh is the earliest evidence yet found for dental work. This evidence comes from the very earliest layers at the site, dating to between 9,000 and 7,500 years ago. As this was the Neolithic era, the tool used to drill the holes would have been made of stone. The authors of a study of teeth from the site suggest that a bow was used to rotate a fine drill bit, which the Neolithic dentists would have wielded to produce a hole in a few seconds. The fact that the holes are limited to the rear molars shows that the goal was not to decorate the teeth—these were not early examples of hip-hop tooth art. They also see evidence of wear around the holes that could have been produced only by chewing after the hole was made, proving that the drilling was done on living people. The implication is that people undertook this painful treatment to try to alleviate the pain of a cavity.

Cavities, extremely rare in earlier, Paleolithic teeth (as well as in modern hunter-gatherers), show a marked increase in Neolithic communities. The best-studied example comes from Clark Spencer Larsen's work on ancient Native Americans. The earliest populations, living a hunter-gatherer lifestyle, have cavities in fewer than 5 percent of the teeth studied. Nearly a quarter of teeth from the period after the adoption of agriculture are afflicted, though—a shocking increase. It's perhaps no wonder that the people of Mehrgarh were willing to resort to a hand-cranked, stone-tipped drill—their teeth were literally rotting out of their mouths!

The increase in cavities during the Neolithic period occurred because the proportion of carbohydrates—starches—in the diet increased dramatically. Paleolithic hunter-gatherers ate a very diverse diet consisting of a wide variety of unprocessed animal and veg-

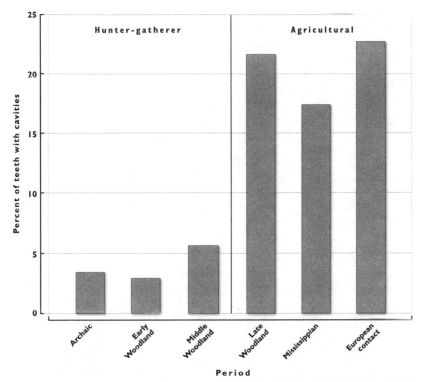

**FIGURE 18: THE INCREASE IN CAVITIES AFTER THE INTRODUCTION OF AGRICULTURE, SEEN AT SEVERAL SITES IN THE SOUTHEASTERN UNITED STATES. THE LATE WOODLAND PERIOD, WHEN AGRICULTURE FIRST APPEARS, DATES FROM A.D. 500–1000.**

etable material, which actually served to clean the teeth during chewing, while much of the Neolithic diet consisted of the processed, starchy seeds of cultivated grasses—separated from the husk, ground, and cooked (often with other ingredients) so that their original identities were entirely lost. I have visited remote villages in the lowlands of New Guinea where nearly all of the calories in the diet come from starchy "puddings" and "pancakes" extracted from the trunks of the sago palm. Not the most likely food source—imagine being the first person to suggest eating the macerated trunk of a rather spiky tropical tree—but a ready source of starch. And it's much easier to make use of abundant starches than it is to gather a wide variety of foods, or to hunt and fish for protein, so people take the easiest route.

FIGURE 19: PREPARING STARCH FROM THE SAGO PALM, KARAWARI RIVER, PAPUA NEW GUINEA.

Today's Atkins diet and its low-carb cousins, including the South Beach Diet and the Paleo Diet, are modified versions of what our ancestors might have eaten prior to the Neolithic Revolution. Atkins perhaps erred too much toward protein in his effort to reduce carbohydrate intake—according to studies of modern hunter-gatherers, foraged plants usually make up the majority of the calories in the diet—but the focus on reducing processed carbohydrates seems to be in line with what we know about what our ancestors would have eaten. It wasn't until the time of the Natufians that grain accounted for a significant fraction of the diet, and that was soon followed by cultivation, as we saw in the previous chapter. Contrast that with the modern diet, where almost all of our calories come from processed carbs and fat, and you can see how far we've strayed from a diet that worked very well for millions of years of hominid evolution.

Of course, starches are better for you than processed sugar. Although sugar extracted from sugarcane, native to New Guinea, has been used by cooks in Asia for thousands of years, the large-scale production of sucrose had to wait until the Industrial Revolution. Before that, Europeans had used honey as a sweetener, and the complex blend of sugars present in honey, as well as its vitamin, mineral, and antioxidant content, made it relatively healthful. It was also fairly rare and quite expensive, and was used as only a very small part of the diet. Industrially produced sucrose, on the other hand, contains nothing but sweet calories; it's perfect for making desserts but adds nothing else to the diet, apart from calories. Industrial production made sugar cheap and plentiful, and soon it was being added to more than desserts.

In *Fast Food Nation,* Eric Schlosser describes the fiendishly complicated science of flavoring processed foods. Chemical additives, natural extracts, and sugar are all added to food products that once would have been consumed *au naturel.* Sugar, in particular, supplements the tastiness of certain otherwise "bland" foods—it is added to chicken nuggets, hamburger buns, and hot dogs in order to make them taste "better." And don't forget the ketchup—one-third of it is sugar. No wonder children love to slather it on everything from French fries to fish sticks. But it's nothing compared to the humble milk shake, which contains massive amounts of sugar as well as a stunning array of added flavorings. "A typical artificial strawberry flavor, like the kind found in a Burger King strawberry milk shake," Schlosser points out, contains these ingredients:

> amyl acetate, amyl butyrate, amyl valerate, anethol, anisyl formate, benzyl acetate, benzyl isobutyrate, butyric acid, cinnamyl isobutyrate, cinnamyl valerate, cognac essential oil, diacetyl, dipropyl

ketone, ethyl acetate, ethyl amylketone, ethyl butyrate, ethyl cinnamate, ethyl heptanoate, ethyl heptylate, ethyl lactate, ethyl methylphenylglycidate, ethyl nitrate, ethyl propionate, ethyl valerate, heliotropin, hydroxyphenyl-2-butanone (10 percent solution in alcohol), $\alpha$-ionone, isobutyl anthranilate, isobutyl butyrate, lemon essential oil, maltol, 4-methylacetophenone, methyl anthranilate, methyl benzoate, methyl cinnamate, methyl heptine carbonate, methyl naphthyl ketone, methyl salicylate, mint essential oil, neroli essential oil, nerolin, neryl isobutyrate, orris butter, phenethyl alcohol, rose, rum ether, $\gamma$-undecalactone, vanillin, and solvent.

Extreme—yes. It's only in the past few decades that science has been able to define the exact mix of artificial ingredients that fool the tongue into thinking it's tasting a strawberry (heaven forbid we add real strawberries!). But sugar is the bigger culprit here, since it's the most common food additive. The reason sugar is so suited to this purpose is that we are genetically programmed to enjoy its flavor. We have a particular taste receptor on our tongues that has been selected over millions of years of evolution to be excited by sucrose, fructose, and other sweet molecules. Why? Because when we were living as hunter-gatherers, long before the local fast-food restaurant existed, one way of telling whether something was good to eat was through its flavor. A bitter taste—which we also have a specific receptor for—meant the possibility of poisonous substances in the food. Think cough syrup or, for those of you old enough, paregoric; chemically intense substances can kill in large doses, so it pays to have a fail-safe mechanism to detect them and prevent us from eating too much of them. But sweet meant safe—think of ripe fruit. By adding sugar, the industrial food producers are tricking our tongues into thinking we're stuffing into our mouths something that's good for us. That's why as children we universally like sweet food but typically develop a taste for bitter foods only much later in life. When was the last time your five-year-old asked for a piece of aged, smelly cheese or a shot of espresso? It takes time to develop the taste sensations that allow us to appreciate more than the dominant bitter flavor in these foods, and to realize that despite millions of years of evolutionary conditioning, they aren't going to kill us.

Sugar, on the other hand, hits us in our evolutionary Achilles' heel, and we can't resist its charms.

The ready availability of processed sugars and starches, coupled with an increase in fat consumption—fat being another evolved taste signal that food is "good"—has raised the number of calories in the typical Western diet by around 15 percent in the past forty years. Add to this a steady decline in the level of activity over the past century as cars have replaced walking and people have moved from jobs requiring physical activity to those that are more sedentary, and it's no surprise that obesity rates have risen nearly tenfold during that time.

Clearly, we are experiencing an epidemic that ultimately has its source in our culture. The industrialization of food production and the reduction in human mobility are relatively recent phenomena, but the diseases that afflict us today are the results of a wave that started back in the Neolithic era. The farmers at Mehrgarh, with their widespread tooth decay, were an early indicator of what would come later. Agriculture—the first form of industrialization, in food production—and, later, mechanical industrialization were unstoppable forces that seemed to take on a life of their own. It's as though agriculture were a virus, expanding in influence despite its negative effects on human health. And it's just begun—diabetes is

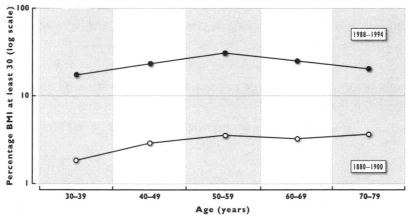

FIGURE 20: INCREASE IN THE PREVALENCE OF OBESITY OVER THE PAST CENTURY. FIGURE COURTESY OF LORENS HELMCHEN, WHO COMPARED UNION ARMY VETERANS OF THE CIVIL WAR TO MODERN AMERICAN MEN. NOTE THE USE OF THE LOGARITHMIC SCALE ON THE VERTICAL AXIS.

projected to be one of the major killers in the United States by
2050, and more than a third of the children born in the year 2000
will develop it in their lifetimes. Add this to hypertension and cancer and you have the three global killers of the future. According to
the World Health Organization, noncommunicable disease will account for more than three-quarters of the global health burden by
2020, up from around half in 1990 and virtually none a few hundred years ago. A wave indeed, and one that is still building.

## BACK TO TENNESSEE

All of this was passing through my mind as I walked around Dollywood in the heat of the Tennessee summer. The utter predictability of where we are headed in the next half century is perhaps
the most surprising aspect. I'm reminded of a scene from the film
*A Fish Called Wanda* where one of the characters is run over by a
steamroller, even though he sees it coming from about a hundred
yards away. He is frozen and apparently unable to move, and the
steamroller slowly rumbles toward him, eventually flattening him
into the pavement. The waves of disease set in motion by the Neolithic Revolution have this same sort of inevitability.

Ultimately, nearly every single major disease affecting modern human populations—whether bacterial, viral, parasitic, or
noncommunicable—has its roots in the mismatch between our
biology and the world we have created since the advent of agriculture. Malaria, influenza, AIDS, diabetes—all could only exist
as significant global scourges in the modern world, with its high
population densities, large populations of domesticated animals,
and high levels of mobility. A sobering fact, and one that should
give us pause as we think about the future we are creating today.

Transgenerational power, the notion that I introduced in the last
chapter, whereby actions taken today can produce effects that last for
many generations, can be a very difficult force to predict, and it's not
always clear what waves will be created and where they will strike.
For instance, who would have predicted that food-preservation
methods using salt—brining and pickling—would have led to an

epidemic of stomach cancer in the early twentieth century, which then, after the invention of refrigeration and a reduction in the consumption of salted food, became one of the rarer cancers in much of the world? Or that the development of the American suburb in the 1950s, intended to create a healthier lifestyle for working- and middle-class families by moving them out of crowded city centers and into open parkland on the urban periphery, would become a contributing factor in the obesity epidemic by producing longer automobile commutes to work? Looking to the future, what will be the long-term effects of having more people over the age of sixty than under the age of twenty, as happened in many western European countries—for the first time in human history—early in the new millennium? An increase in diseases of the elderly, surely, but also perhaps social unrest as the pension crisis hits and the shrinking pool of young workers is forced to bear the burden of caring for the older generation. Ultimately, though, these are relatively short-lived trends, playing out over the course of a couple of generations. What about trends that take place over hundreds or thousands of years— how can we make predictions about these?

What, for instance, are the long-term trends in human disease? We will look at some possibilities in Chapter 5, but for now we need to consider another growing trend: a different sort of disease, but one that is increasing at rates similar to that of diabetes. According to the WHO, mental illness will be the second biggest cause of death and disability by 2020. Yes, the *second* biggest, superseded only by heart disease. Over 400 million people around the world are afflicted by illnesses ranging from epilepsy to schizophrenia to depression. Suicide rates are climbing; over 1 million people every year take their own lives, more than are murdered or killed in wars. And which countries are leading the way, with rates of depression and anxiety that are unrivaled in the rest of the developed world? America and Japan—perhaps the most technologically advanced societies on the planet, and among the richest. Clearly, technology and affluence alone don't make people happy. The alarming trend in declining mental health is what we'll look at next.

# Chapter Four
# Demented

Few topographical boundaries can rival the frontiers of the mind.
—SALMAN RUSHDIE,
*Imaginary Homelands*

We all face the same way, still it takes all day.
—THE STEREOPHONICS,
"Traffic"

## MARIA GUGGING, AUSTRIA

I stepped off line 4 of Vienna's U-Bahn at the Heiligenstadt terminus and walked upstairs, out into the chilly November air. This underground railway stop is a busy transportation hub, with buses connecting the city to outlying areas, and commuters to-ing and fro-ing in the gray morning light. I was immediately surrounded by people rushing to appointments or the shops, tugging coat collars up around their necks to keep warm. I was not here for a business meeting or to run an errand, though. Rather, I was headed into the northern Vienna suburbs on a somewhat unusual quest. After locating the right bus stand, I hopped aboard and paid, confirming with the driver in my rusty college German that the bus stopped where I needed to go.

The ride out of the city gave me time to emerge gradually from the touristy bustle of Mozart, Sacher tortes, and Viennese commuters. As the old Vienna Woods surrounded me and the villages rolled past—Kammerjoch, Leopoldsbrücke, and Klosterneuburg, with its massive castle looming over the road—I felt as though I was traveling back in time. After about half an hour I reached my destination, a small village named Maria Gugging.

The name Gugging is emblazoned in the mind of every Austrian, but not, perhaps, for the best of reasons. This is because Gugging is pretty much a one-shop town, like Cupertino, Anaheim, and Nashville. But unlike the hometowns of Apple, Disneyland, and country music, the business in Gugging is the human psyche. Maria Gugging is the home of the Lower Austrian Psychiatric Hospital. The hospital itself still evokes memories of an earlier, not-quite-so-enlightened time in the history of psychiatry—a time when there were hundreds of patients, sleeping fifteen or more to a room, with bars over the windows and only two doctors on staff. During the Nazi era, and in the early post–Second World War days, the aim was to remove "insane" people from the general population, to lock them away where they couldn't do any harm to themselves or others. This goal has changed over the years as more sympathetic doctors have taken over from their less enlightened forebears. But walking the footpaths among the gray institutional buildings, especially on a cloudy late-fall day, I had a vague sense that the ghosts of lonely, tortured patients were watching me.

It was along one of these paths, about half a mile from the main gate, that I approached a smaller building nestled at the edge of the forest. I could tell that I'd reached someplace extraordinary because, uniquely among the buildings at Gugging, this one was covered in multicolored abstract art—a face here, a splash of seemingly random graffiti there, all rendered in a phantasmagoric, colorful style that created a vaguely childlike impression. It was this art, and the people who'd created it, that I was here to see.

Upon entering the building, accompanied by my guide, Nina Katschnig, I was confronted by one of the residents, who asked me for a cigarette. "They smoke too much, but what can we do?" she said, shrugging her shoulders. The man seemed disconnected from his surroundings, wrapped up in his own world, from which he emerged occasionally to exclaim or to ask a pointed question in guttural Austrian German. We walked on and entered a small kitchen area, where Nina introduced me to Dr. Johann Feilacher, a psychiatrist and artist, and the director of the building I was now standing in.

The Haus der Kunstler, or House of Artists, was created in 1981 as the Center for Art and Psychotherapy. The first residents at the center were eighteen patients from the hospital down the hill, a collection of schizophrenics, psychotics, and manic-depressives united by one thing: they were all very good artists. The goal of the center was to apply the theories of its founder, Dr. Leo Navratil. Navratil had come to the hospital as a psychiatrist in the 1950s, and during the late 1950s and early 1960s he began experimenting with art therapy, encouraging his patients to draw and paint as a form of treatment. After reading the books of the famous French surrealist artist Jean Dubuffet about the untrained art of children and the insane, Navratil contacted him about his own ongoing therapeutic work at Gugging. Encouraged by Dubuffet, he wrote the influential book *Schizophrenia and Art,* published in 1965.

Navratil thought that the minds of schizophrenics and other mentally ill people could provide an insight into the artistic process. He argued that schizophrenics are uniquely in touch with the font of creativity in a raw, untutored way that allows them to create "pure" art. According to Dr. Feilacher, who took over as director of the House of Artists in the early 1990s, after Navratil's retirement, "Navratil believed that psychosis makes art—that every psychotic man or woman is an artist." Navratil was clearly a man of his time, the 1960s and '70s, with its emphasis on personal psychedelic journeys. The message that psychosis (whether drug-induced or not) can create art had an eager audience. The first public exhibition by the Gugging artists was in 1970, and after this their work became well known to the entire art world.

What actually goes on inside the heads of the mentally ill artists in Gugging is still the subject of much scientific debate. Some, though certainly not all, are schizophrenics. Navratil noted that some of them had low intelligence, often caused by brain damage in early childhood. Many had experienced extremely difficult upbringings, with poor educations and a history of neglect. This mistreatment is certainly a significant influence on their art, some of which is quite violent.

August Walla is an especially interesting resident, since he is one of the most famous of the Gugging artists. Pieces of his work have sold for more than $10,000 and can be found in collections around the world. Walla lived with his disturbed mother and grandmother in a garden shed near the Danube for many years, collecting discarded items from the surrounding countryside and decorating the inside of the small building with fantastic paintings. There were never any male figures in his life, a conscious effort on the part of his mother to keep him isolated; as Feilacher explains, "His mother kept him as a child," perhaps fearing for his safety if he strayed beyond the family's odd, insular existence.

They lived in an open space of only about fifty square feet, with a cooking stove in the center surrounded by a mountain of rubbish. When Walla entered the House of Artists because his mother could no longer take care of him, she was admitted to the hospital down the hill. For his first ten years there, he refused to speak to any of the other residents, or to Dr. Feilacher, conversing only with his mother—a continuation of his fifty years of nearly total isolation from the rest of the world while living in the shed. During this

**FIGURE 21:** *ADAM AND EVE IN PARADISE* **BY AUGUST WALLA.**

time he also wrote on trees and decorated every space in his room—
and eventually the outside walls of the building—with fantastic
paintings, in a manic outpouring of creative energy. He remained,
however, completely wrapped up in his insular family universe.

When his mother died, at the age of ninety-seven, this changed
abruptly. Suddenly Walla emerged from his shell and spent all day,
every day, talking to anyone who would listen. He followed Dr.
Feilacher around like a child for months, hounding him with an
endless disjointed monologue. In effect, Dr. Feilacher, and then the
rest of the inhabitants, replaced his mother. This manic phase,
gradually less extreme as time wore on, lasted nearly ten years, until his death in 2001. Interestingly, during this part of his life
Walla produced very little art of any substance—all of his great
work stems from the earlier, insular period, as though the only way
he could communicate with the world beyond his mother was by
painting and drawing. Navratil felt that untrained artists such as
Walla, cut off as they are from the rest of the world and oblivious to
any notions of artistic training or concern for prevailing fashions,
give an insight into the raw creative process. In effect, he said, insane people show us what we are all capable of if we can only throw
off society's constraints.

After twenty-five years of working at the House of Artists, Feilacher dismisses Navratil's notion that all insane people are artists;
he believes that the incidence of artistic talent is the same among the
inhabitants of mental institutions as it is in the outside world. "As
an artist myself," he told me, "I see them as colleagues"—people
who, although certainly psychologically lacking in some ways, clearly
have a gift that deserves to be shared with the rest of the world.
Feilacher also told me that although he finds much of Navratil's
thinking simplistic, schizophrenics do possess an interesting trait:
"What you can see is a return to roots—by this I mean a return to
childhood." But what is it about this childlike state that engenders
art? Children, of course, feel compelled to draw, to mold clay, to
fold paper; the raw artistic process rises from deep within their own
psyches. But why?

The art created by Walla and the other members of the house has come to be called "outsider art," or *art brut,* to use the term originally coined by Dubuffet. This type of art is produced by people unschooled in a particular artistic genre, and it is often considered to be part of the primitivist movement. If, goes this school of thought, art has become so corrupted by formal movements, trendiness, and marketing, then it is only by returning to the work of "primitive" cultures, children, and the insane that we can gain some sense of the purity of artistic expression.

In keeping with this "purity," the artists at Gugging are not very interested in the work of other artists. A Swiss television crew making a documentary on the house once took a group of them to a local museum to see how they would react to the art there, but they ignored the paintings on the walls. "The most interesting place in the museum was the coffee shop," Feilacher told me, laughing. The best art to them is their own, and they live wrapped up in worlds of their own imagining. How can such people create beautiful artistic works when they are largely unschooled in the intricacies of art history, style, technique—the preconditions to being a proper artist, according to most art teachers? And what drives them to create in the first place?

I came away from Gugging with an insight into the raw creative process and a desire to understand more about what impels us to take images from the world around us—or from inside our own heads—and re-create them in two- or three-dimensional representations. Did prehistoric, untrained artists—like the residents of the house—create art out of a compulsion? Perhaps. But assuming that human behavior is adaptive, it's very difficult to explain the existence of art for art's sake. After all, you aren't going to bring down many gazelles or outrun a pack of wolves by painting a picture of one on the wall of a cave. What does art—even less refined art, such as that produced tens of thousands of years ago—reveal about the inner working of our minds and the dawn of human consciousness? For this we'll need to take a short detour through modern molecular genetics, en route to the African savannas of 70,000 years ago.

## A SPEECH IMPEDIMENT

In 1996, a group of physicians at the Institute for Child Health in London approached Tony Monaco, a professor of human genetics at Oxford University, with an interesting case study. The physicians had been studying a group of relatives that, to preserve their privacy, they called "the KE family," a family of Pakistani origin with an inherited speech impediment. Members of the family going back three generations lacked the ability to articulate words because they could not control the necessary movements in the lower half of their faces. They also had problems with grammar and generally couldn't make themselves understood to outsiders.

Interested, Professor Monaco and his team conducted what is known as a "genome scan"—an analysis of hundreds of variable locations in the genomes of KE family members with and without the speech disorder. The idea was that if a particular set of these variable markers tagging a specific genomic region was consistently found in the affected family members but *not* in the unaffected relatives, it would be likely that the genetic change that led to the disorder would be found somewhere in that part of the genome. After a year of painstaking work, they found a region on chromosome 7 that was associated with the speech defect. The problem was, there were around seventy known genes in the region, and narrowing down which one was responsible would not be easy.

Then they got a stroke of good luck. Independently, a physician in Oxford encountered an unrelated child with a speech disorder that sounded suspiciously similar to what had been described in the KE family. This individual, known as CS, was analyzed by Monaco's team. The results revealed that this patient had a chromosomal rearrangement known as a translocation, where one part of a chromosome is cut and spliced into another chromosome. When this happens and the break point is in the middle of a gene, the function of the gene can be disrupted. The gene that was disrupted in CS, known as *FOXP2,* was also mutated in the KE family. This was the first time that a change in a single gene had been shown to have an effect on speech, and the publication of this discovery in the prestigious scientific journal *Nature* in 2001 received enormous fanfare. At last, said some journalists, the "language gene" had been discovered.

*FOXP2* stands for *F*orkhead b*ox* protein *P2,* and it is a member of a class of proteins known as transcription factors. The proteins interact with the DNA in such a way that they turn other genes on and off—they are like the "molecular coaches" of the genome, substituting players and calling the shots in the game of genomic function. It is because of their effects on the way many other genes are turned on or off that changes in these transcription factors can have complex effects on one's physical and mental characteristics. This helps to explain why mutations in this single gene can affect something as complex as speech and grammar, which were previously thought to be controlled by hundreds of genes. Because of this central role it plays in gene regulation, *FOXP2* has been highly conserved in evolution; very similar forms are found in chimpanzees and mice. Interestingly, when the gene is mutated in mice they exhibit signs of a speech disorder, producing improper vocalizations as babies. While mice clearly don't have the sort of complex spoken language that humans do, the result suggests that a similar function has been preserved during over 70 million years of evolution.

The *FOXP2* results immediately raised the question of whether changes in its structure could have been one of the key biological changes that allowed our human ancestors to develop speech. The extent to which other hominids—australopithecines, *Homo habilis, Homo erectus,* and Neanderthals—were capable of communicating with one another has been one of the most hotly debated subjects in physical anthropology. Most researchers believe that the earliest species, the australopithecines, had rudimentary language skills, not unlike those of chimpanzees. As the brain enlarged during evolution, first in *Homo habilis* and then even more so in *Homo erectus,* language skills probably became more complex. By the time of the Neanderthals, who split from the lineage leading to *Homo sapiens* around 500,000 years ago, it's thought that spoken language had appeared. This is supported by a wonderfully preserved 60,000-year-old Neanderthal skeleton from Kebara Cave in Israel that has an intact hyoid bone. The hyoid is the delicate bone in your throat that provides the structure for the Adam's apple, and it helps us to modulate spoken sounds. The fact that Neanderthals had a hyoid bone similar to humans' suggests that they, too, may have been capable of complex speech. But what did their *FOXP2* gene look like—did they have the genetic capacity for language as well?

Unfortunately, there are no living Neanderthals to sample, or finding the answer would be easy. In some cases, though, it's possible to analyze DNA from long-dead specimens. The field of ancient DNA research is to genetics what high-stakes poker is to gambling: there are high risks, but also high rewards. DNA is not a terribly stable molecule, and it usually degrades soon after death. Because of this, most attempts to extract DNA from ancient remains end in failure. Yet in extremely rare cases, it is possible to obtain intact DNA from specimens that are even tens of thousands of years old. This is typically possible only when the sample has been preserved in cold, dry conditions—such as those in European caves. And it is just such a cave that has yielded DNA from a very rare specimen indeed.

Vindija Cave, in northern Croatia, has been studied by archaeologists for over a century. It contains hominid remains ranging in age from 25,000 to 45,000 years old—a very important period in European prehistory. It was during this time that Neanderthals were replaced by modern migrants recently arrived from Africa via the Middle East. The deepest—and thus the oldest—layers in the cave contain only Neanderthal material, while the layers at the top are composed entirely of modern human remains. One of the Neanderthal bones from the deeper layers yielded intact DNA, and employing the painstaking methodology necessary to tease results from ancient DNA, scientists managed to slowly assemble the sequence of the *FOXP2* gene. The publication of this result by Svante Pääbo and his team in October 2007 shocked many who had taken the earlier *FOXP2* results as evidence that just a few changes in this gene could have led to fully modern language. The Neanderthal specimen had the human form of the gene!

The reason this came as such a shock was that an analysis of *FOXP2* carried out by Pääbo's group five years earlier had suggested that the human form had arisen much more recently. There are only two differences between the amino acid sequences—the protein building blocks encoded by the gene—of the *FOXP2* genes in humans and chimpanzees. Only two differences, out of the 715 amino acids in the protein, even though humans and chimps last shared a common ancestor around 5 million years ago. Moreover, from the genetic patterns seen in human populations, Pääbo and his group concluded that these changes had appeared within only the past 200,000 years. This result suggested that modern humans alone were defined by the unique *FOXP2* sequence that conferred language abilities. The newly obtained Neanderthal sequence, though, showed the earlier interpretation to be wrong. The evidence from *FOXP2* was that our distant hominid cousins had also had the capacity for spoken language. What did it mean—had Neanderthals really spoken the way we do?

The answer is almost certainly no. Our burly cousins were very similar to modern humans in most ways, yet in several key characteristics they were quite different. As I mentioned before, the Neanderthal and modern human lineages appear to have split from each other around 500,000 years ago, based on their levels of DNA divergence. Around this time, the ancestors of the Neanderthals, likely belonging to a species known as *Homo heidelbergensis,* left Africa and moved into Europe, while our ancestors stayed in Africa. Over time *H. heidelbergensis* evolved into *H. neanderthalensis,* a species whose heavyset bodies were wonderfully adapted to the cold climate of Europe. Fully Neanderthal physical characteristics appeared around 150,000 years ago, and these bulkier hominids later expanded their range into western and central Asia during the last ice age. For around 100,000 years they were the masters of the continent. Then we appeared on the scene, starting with an African exodus between 60,000 and 50,000 years ago. We entered Europe around 35,000 years ago, and within a few thousand years of our appearance on the scene, the Neanderthals were extinct. The reasons are still hotly debated, but a consensus is beginning to form.

Thirty-five thousand years ago, the world was in the firm grip of the last ice age. In fact, there was a strong cooling trend around that time that caused the forested landscape of Europe to be replaced by open grassland and tundra. Neanderthals, with their simple technology and emphasis on brute-force methods of hunting, would have been at a disadvantage in such an environment. Their hunting methods worked fairly well in wooded countryside, where they could hide and pounce on prey at close range, but they faltered in a more open landscape. Conversely, the larger group sizes, highly social hunting structure, and more sophisticated tools of modern humans gave them a strong advantage when hunting in open land. Relative to the comparatively sophisticated newcomers, Neanderthals were yesterday's news.

What gave modern humans such an advantage? Clearly, two factors were working in our favor in Europe: climate change and social sophistication. But while climate change was beyond our control (unlike today), the highly social nature of modern humans was determined by internal forces. What was it that made us such social beings, and when did this change occur? It was certainly after we diverged from Neanderthals, but when was it, and what caused it?

Deciphering what this change was is perhaps the one research topic in anthropology that will win its discoverer a Nobel Prize (or at least it should). For decades researchers have been debating this key change on the road to making our species fully modern, and there are many theories. All seek to explain the remarkable transformation in human social behavior manifested through an enormous change in toolmaking styles at the beginning of the Upper Paleolithic period, or the Late Stone Age, as it is called in African prehistory. Around 60,000 years ago, tools start to become much more finely crafted, and there is evidence for the use of bone in shaping extremely fine spear points—something that no hominid had done before. Beginning around that same time, there is also evidence for an expansion in population, which led to the settlement of the world outside of Africa. Only 40,000 years later *Homo sapiens,* a species that had been limited to Africa during the first 150,000 years of its time on earth, had expanded to nearly every habitable location on the globe, settling Asia, Europe, Australia, and the Americas in only around fifteen hundred human generations.

I discussed this global march in *The Journey of Man* and therefore won't revisit it in detail here. Suffice it to say, it's a story of endurance and ingenuity writ large—*The Odyssey* of our species. As we moved from our homeland in tropical Africa into environments as diverse as the mountains of central Asia, the valleys and caves of southwestern Europe, the tundra of Siberia, and the jungles of Amazonia, we adapted in many ways to the new geographies. No primate species had ever been able to expand its range to such an extent, and we were the master of all we surveyed. Consummate hunters, with highly adapted tools and advanced hunting skills, we made short shrift of the Neanderthals in Europe. But what gave us such a huge advantage compared to every species that had come before on the hominid line?

A better brain, it turns out—the triumphant march of humans across the globe was preceded by a march of neurons inside our skulls. An average human brain is around 1,400 cubic centimeters, while our earliest hominid ancestors, living around 4 million years ago, appear to have had apelike brains of around 500 cubic centimeters. Despite this, they walked upright—an adaptation to life on the tropical African savanna, where reducing the surface area exposed to sunlight, seeing farther, and moving more efficiently would have been favored. The earliest members of our own genus, *Homo habilis* at 2.4 million years ago and *Homo erectus* at around 1.8 million years ago, had brains on the order of 750 and 1,000 cubic centimeters, respectively. Along each step in this progression the tools became more sophisticated, archaeological evidence for the increasing encephalization occurring under the hood. Then things get a bit strange. The average Neanderthal, it turns out, had a brain of around 1,500 cubic centimeters—larger than our own, on average. Why then, you might be wondering, didn't the Neanderthals, armed with large brains, hyoid bones, and a humanlike version of *FOXP2,* manage to fight off the *Homo sapiens* onslaught in Europe?

Clearly, size isn't everything—it's how you use it that counts. The Neanderthals were indeed more sophisticated than their ancestors with smaller brains, and were actually quite successful for their time. And their *FOXP2* sequence and hyoid bones hint at a capacity for language. What seems likely, though, is that if they could talk, they didn't have much to say. Neanderthals were probably pretty boring. And the reason we know this is that they didn't have art, which made its appearance only in the past 70,000 years, and then only in our species.

Think about it: we are not the only species capable of communicating, although most communication occurs in ways that don't involve speech. Every creature, from corals capable of distinguishing between "me" and "you" on the basis of chemical cues, to dogs sending nonverbal signals to each other to reinforce their social hierarchy, to frogs that sing in the darkness of a tropical forest to locate their mates—all are communicating, transmitting understandable signals to one another. We're not even the only species capable of verbal communication; those same frogs, as well as birds and whales, "talk" to each other. But what do they say? Clearly quite a bit that is necessary for their survival and reproductive success, which is how their communication systems evolved in the first place. A deaf meerkat, unable to hear his colony-mate on sentry duty signaling at the approach of a predator, would be unlikely to live long enough to leave offspring. Communication is not uniquely human by any means.

But humans are not created equal; there is tremendous variation in our innate abilities. Most of us are capable of basic "human" tasks—learning a language, solving rudimentary problems, mak-

ing simple tools, and so on—but there is a range of variation for all
of these traits. Think of Picasso or Einstein, Shakespeare or da
Vinci. Not all of us can create extraordinary art or design complex
machines, but we can make use of the insights of others, because of
our sophisticated social structure. The transmission of ideas in
modern human populations is unprecedented in evolutionary history. The *Homo sapiens* archaeological record is, in effect, a story of
one innovation triumphing over another, the march of ever-better
mousetraps through time and across space. The Upper Paleolithic
population expansion was the result of such innovation, when we
developed the adaptable mind that would allow us to take on the
challenges of a global empire.

Complex speech allowed early humans to communicate new
ways of solving the problems that came their way. If our species
had been highly adapted to a single ecological zone, such as the
tropical forest or the high mountains of eastern Africa, it is likely
that we never would have developed the complex minds we possess
today. In fact, we have always been a species living in environmental flux. The savannas of Africa, with their complex mix of grasses,
trees, and extreme seasonal changes, were the perfect breeding
ground for adaptability. As the climate changed over the past few
hundred thousand years, the shifts in the extent of the savanna encouraged population expansion, followed by contraction as the
changing climate reduced the size of the environmental resource
base. This led some early humans to live by the sea, eating the
plentiful shellfish, as early as 120,000 years ago. It also drove some
to seek greener pastures in other places.

As with the dawn of the Neolithic era, our migrations out of
Africa and throughout the rest of the world were probably set in
motion by necessity—in this case, dwindling supplies of food and
water. There is evidence for a small population migration around
110,000 years ago, when *Homo sapiens* skeletons turn up in the
Middle East. Data from African lake sediments suggests that the
continent was becoming much drier around this time, and this dry
spell might have encouraged early human populations to migrate

in order to find the food resources they needed. Ultimately, these pioneers disappear in the Middle East after 80,000 years ago, replaced by the cold-adapted Neanderthals at cave sites such as Qafzeh and Skhul. Why? It was around this time that the ice age really began to get rolling in a major way—it would eventually cover most of North America in ice. And this cooling process seems to have been kick-started by a wrench that was thrown into the global climate works around this time.

## VOLCANOES AND MACROMUTATIONS

Today, Lake Toba, in northern Sumatra, is one of the jewels of western Indonesia. From the air it looks like a beautiful blue oval ring of water, around sixty miles long, with a large green island in the center. Its appearance seems somewhat odd until you learn how it was formed. Lake Toba is actually the water-filled caldera of a dormant volcano, known as Mount Toba, and is much like Crater Lake in Oregon—except Toba is ten times larger. In fact, it is the largest volcanic lake in the world, which means that the volcano it sits in must have been quite big as well. Toba was one of the largest volcanoes ever, and when it last roared to life its eruption was the largest in the past 2 million years, spewing out nearly three thousand times as much material as the eruption of Mount Saint Helens in 1980. And when did this extraordinary calamity happen? Sometime between 75,000 and 70,000 years ago.

When volcanoes such as Toba erupt—and it isn't very often—the material they spew into the atmosphere does more than simply blanket the surrounding region with a layer of ash (parts of central India were apparently buried under eighteen feet of Toba's). The force of the eruption is such that much of the material, as well as the sulfuric acid from the blast, is forced into the higher layers of the atmosphere, where strong winds disperse it around the world. Since Toba is near the equator, its ash would have been dispersed quite easily throughout the region that receives the most sunlight: the tropics. The result was an instantaneous cooling effect, due to

the particle-obscured sunlight, that lasted for several years—
a "volcanic winter," in which global temperatures dropped somewhere between nine and twenty-seven degrees Fahrenheit. This
brief episode, devastating though it must have been, happened to
then be followed by approximately 1,000 years of substantially
cooler temperatures, among the coldest of the last ice age. Africa,
as the most tropical continent on earth (85 percent of it lies between the Tropics of Cancer and Capricorn), would have experienced this cooling, but it would also have become drier as more of
the earth's moisture was tied up in expanding ice sheets. Overall,
the effects of the Toba eruption, coupled with the sudden onset of
an extremely cold period during the last ice age, would have been
devastating to early human populations.

Genetic evidence suggests that around this time the total number of people alive fell to fewer than ten thousand—perhaps, as I
mentioned in Chapter 1, to as few as two thousand, according to a
recent paper by the geneticist Marcus Feldman and his colleagues
at Stanford University. The genetic and climatic data both paint a
picture of a human population teetering on the brink of extinction.
It is likely that the cataclysmic climatic shift created a scenario in
which humanity had to adapt or die. And the response of these
humans—as with the response of the Natufians in the Middle East
60,000 years later—was to change their culture.

It began slowly, with mere hints of the extraordinary changes to
come. The first pieces of evidence show up around the time of the
Toba eruption, perhaps even a bit before. Beginning around 75,000
years ago, pieces of soft stone known as ochre, carved by human
hands with complex geometric motifs, start to appear in the archaeological record. While obviously not the work of a precocious
Michelangelo, they are the first evidence of art in the history of our
(or any other) species. The creation of a clearly decorative motif on
a stone that may have been used as some sort of primitive counting
tool represents a distinct break with the culture of our hominid ancestors.

Many readers will have already heard about the "Great Leap For-

FIGURE 22: THE WORLD'S EARLIEST ARTWORK. ENGRAVED OCHRE FROM BLOMBOS
CAVE IN SOUTH AFRICA, DATING TO AROUND 75,000 YEARS AGO.

ward," as Jared Diamond has called the abrupt change in behavior
that heralds the Upper Paleolithic period, or the Late Stone Age.
For many years archaeologists believed that these early signs of
fully modern human behavior—art, finely crafted stone tools, advanced hunting techniques—appeared only between 50,000 and
40,000 years ago. With the new discoveries of decorative African
art dating back more than 70,000 years, what now seems clear is
that these younger archaeological finds mark the expansion of people who had been refining their culture for tens of millennia prior
to this. Instead of having to account for a sudden change in human
behavior that explodes on the scene within a few thousand years of
the Upper Paleolithic, probably caused by a few genetic mutations,
judging from the suddenness with which they appeared, archaeologists now increasingly believe that the process was much more
gradual. Humans, it seems, were probably preadapted to develop
the material culture of the Upper Paleolithic period, and all they
needed was the impetus—in the form of the intense selective pressure provided by the last ice age and the eruption of Mount Toba—
to make use of their ability to solve problems in novel ways.

The old model of a genetic "revolution" leading to the Upper
Paleolithic period was influenced by the work of Richard Goldschmidt, an important twentieth-century German-born American

geneticist. Research on gypsy moth evolution and sex determination led Goldschmidt to question the Darwinian model of how species evolve. This is actually one of the least understood processes in evolutionary biology, one even Darwin recognized as a challenge for his theory of evolution by natural selection. How does one species suddenly become two? Traditional Darwinian theory argues that this process—known to evolutionary biologists as macroevolution—follows the same process as change within a species, which is known as microevolution. If two populations of the same species become geographically isolated from each other, it is easy to see how, over time, small changes could eventually lead to enough genetic divergence that the populations become separate species. However, much of macroevolution seems to have taken place without such geographic isolation, through a process known as sympatric speciation, where the daughter species aren't separated by geographic barriers. How might this occur?

Goldschmidt proposed the idea of "macromutations"—genetic changes of large effect that create what he called "hopeful monsters." In this model, he argued, not all mutations are created equal; some have a larger effect on the organism than others. For instance, some plants have formed separate species by doubling their chromosome number—the genetic process that led to the polyploid wheat and corn genomes we learned about in Chapter 2. In these cases the vastly different chromosome numbers in the plants prevent them from interbreeding, so two species are effectively formed instantaneously. Although this appears to have been a relatively common way for plants to speciate, it is unlikely to have been widespread in animals because they are generally incapable of reproducing through fusing sperm and eggs produced by the same organism (a process known as *selfing*). If you don't have anyone who is reproductively compatible with you, your lineage dies out. But perhaps the macromutations envisioned by Goldschmidt weren't this extreme, and they only made reproducing with members of the original population more difficult. The bearers of such genetic changes—the "hopeful monsters"—might be able to produce enough offspring to found a new population, one in which reproductive compatibility was normal among its own members. Selection would then act to discourage attempts at interbreeding, and the populations would be on their way to becoming separate species.

While the utility of polyploidization for explaining the sympatric speciation process in plants is straightforward, we need another type of macromutation for sexually reproducing animals. It is certainly true, though, that seemingly trivial changes in an organism's DNA can have an enormous effect on its physiology or behavior. Sickle-cell anemia, a devastating disease common in West Africa, is caused by a single nucleotide change. And of course *FOXP2,* as we have just seen, can destroy someone's ability to speak with a single mutation. Was the change that first allowed our ancestors to speak an example of such a macromutation? Possibly, although as we just learned, Neanderthals had the human form of the gene and yet left no archaeological evidence that they were capable of complex thought and speech. How, then, did the behaviors that define the Upper Paleolithic era—and fully modern human behavior—come about?

In a 2003 paper in the scientific journal *Nature,* the American evolutionary geneticist Richard Lenski and his colleagues suggested a model for how this transition might have occurred. The paper described a computer program they called Avida and its utility in explaining the evolution of complex traits such as eyes. Darwin had described these as "organs of extreme perfection and complication," and their existence seems difficult to explain using microevolutionary processes—what use is 10 percent of an eye, after all? Avida modeled the evolution of digital organisms— sequences of 0's and 1's—over many generations of mutation, reproduction, and natural selection. Starting from an identical pool of organisms, the researchers let Avida's computationally simulated evolution take its course, and gave "rewards" to their digital menagerie based on how well its members performed certain preassigned evolved computational functions in the digital universe, with more complex functions given higher rewards. The most complex of these, known as EQU, required a minimum of five "mutations" from the ancestral state, but of course you would have to know exactly what you were mutating toward in order to achieve this in only five steps. As in nature, Avida's mutations were random, so the actual number of generations and steps was far higher than in the simulations.

What Lenski and his coauthors showed in this simple computer experiment was that extremely complex traits—such as EQU, the abstract computer-code equivalent of an eye—could evolve from preexisting changes that occurred in an organism's evolutionary past. Macromutations do occur—and, in fact, Lenski and his colleagues did see examples of multistep changes in their computer model—but they weren't necessary to explain the evolution of a complex, multifaceted trait like EQU. Rather, it was the gradual accumulation of a few such changes of small effect, in the *right combination* and with strong *natural selection,* that eventually created the trait. What this showed was that microevolution could explain traits like eyes after all, and thus in the end Darwin was right.

The obvious implication for our story is that a complex modern human trait like the capacity for abstract thought, first recognizable in the fossil record through artistic depictions, could have arisen in the same way, through small, incremental steps that eventually led to the right combination for natural selection to act on. This model might explain why we see sporadic evidence of modernity prior to 70,000 years ago, but only see it explode afterward. The individual genetic mutations that made such behavior possible had existed for perhaps tens of thousands of years, but their *combination* in such a way as to produce abstract thought was strongly selected for only after this time. The extreme climatic changes brought about by the ice age and, in all likelihood, the eruption of Mount Toba would have exerted on the human species selection for innovation and speedy adaptation—such strong selection, in fact, that we developed a new culture. Like the dawn of the Neolithic era 60,000 years later, a climatic crisis paved the way for cultural innovation.

The fact that the same skills that allowed us to be more effective hunters and gatherers also gave us the eventual ability to write sonnets and compose electronic music is perhaps not such a mystery after all. What evolved around 70,000 years ago in the human lineage was the ability to adapt quickly—to innovate—using our *culture,* as opposed to our *biology,* as was the case with Neanderthals. Innovation is a complex process, but at its most basic it involves imagining new ways of solving a problem and then implementing them. The first step requires the sort of imagination that is reflected in the creation of art, like that produced by the untrained Gugging artists, and the second requires some way of explaining the innovation to others. The very process of imagining new possibilities is accelerated by cross-fertilization, in the same way that early agriculturalists crossbred different strains of wheat, rice, and corn from the fragmented, mountainous habitat in which the plants had originally evolved in order to create the traits they wanted. This process of trial and error (often using seemingly crazy insights) coupled with better communication became the model for human innovation—the first time such a successful system for problem solving had ever evolved. The change in human behavior at the dawn of the Upper Paleolithic period can be explained only by these two abilities working in concert.

Modern hunter-gatherers provide a wonderful case study of this behavior. Everyone sits around the fire at the end of the day, telling stories, laughing, and discussing the day's events. Some of these stories become part of group mythology—an account of a particularly successful hunt, perhaps—while some are a way of testing and refining new ideas. It's a sort of "innovation think tank," with the members of the group performing thought experiments as they discuss, dissect, and decide on narratives about their lives. This process of narrative refinement is not unlike the internal cellular process by which our short-term memories are turned into long-term ones—through reiteration of the story, reinforcing the neural connections that tell the narrative in the way we *want* it to be told. Modern humans, in effect, evolved to be social machines that produce and refine ideas, and perhaps this explains why management studies show that people seem to work best in the kind of small, focused teams where this ancient hunter-gatherer process can take place.

This model for the evolution of a rapid innovation helps to explain why the Neanderthals were doomed from the moment modern behavior appeared. The changes that led us to innovate also made us curious, probably created a sense of wanderlust, and enabled us to change our culture incredibly quickly in response to new conditions. In effect, the seeds of the Neolithic era were really sown 70,000 years ago; it was just a matter of finding the right set of conditions in which the next step could occur, which would only come when the climatic conditions were right and human population densities had increased to the point where hunting and gathering were no longer viable alternatives. In this model, art and other nonadaptive accoutrements of modern behavior are what evolutionary biologists Stephen Jay Gould and Richard Lewontin would have called *spandrels*—by-products of other evolutionary forces, and not necessarily ends in themselves.

There is a problem with the success of our cultural adaptability, though. In the process of creating a densely populated, agricultural way of life, we were forced to subsume our individual desires for novelty to the desires of the broader culture. In contrast to our hunter-gatherer ancestors, who were free to explore any and all cultural possibilities, from fishing for salmon to hunting mammoths on the central Asian steppes to creating beautiful artistic depictions on the walls of French caves, their Neolithic descendants were forced to channel theirs in order to suit the broader needs of society. In effect, minds that had once been free, with the endless territory of the Paleolithic globe in which to realize their musings, were now caged, limited in both geography and focus. We had gone from living in "the original affluent society," as anthropologist Marshall Sahlins famously referred to hunter-gatherer populations, with free to time to devote to seemingly idle activities, to being a group of worker bees with looming deadlines to meet.

This process of behavioral specialization was codified in the Hindu caste system of India, in the hierarchy of the Catholic church, in the rigid Confucian meritocracy of China, and in the feudal systems of medieval Europe. Society achieved its goals only if the individual components worked together in intricate harmony— a cog couldn't strive to become a bolt or the whole machine would seize up. The process accelerated during the Industrial Revolution, as the benefits of specialization became even more apparent. In effect, people started to merge with their machines, spending their whole lives performing repetitive tasks that, while wonderful at producing large quantities of standardized, inexpensive goods, effectively robbed the average factory worker of his or her individuality and creativity.

Even in today's postindustrial countries, the process of specialization continues, with subcultures in the same society speaking entirely different languages during most of their waking hours. Would the average tort lawyer understand a seminar in subatomic physics? How many chemists would be able to contribute meaningfully to an academic conference on literary theory? And when was the last time you worked on your own car? We tolerate this be-

FIGURE 23: SCENE FROM THE TUCKERS FACTORY IN BIRMINGHAM, ENGLAND, 1910. (PHOTO COURTESY OF PAUL TUCKER.)

cause of the benefits it provides—from the point of view of society, it is better to have experts focus on a limited aspect of human endeavor and excel than to have generalists spend their time as dilettantes, moving from one task to another as their interests suit them. While this has produced an obvious bounty in the form of advanced technology, for instance, it has left many people feeling out in the cold, excluded from the decision-making process about their own lives.

In the final chapter of this book I will discuss how this alienation has been a driving factor in the rise of fundamentalism during the late twentieth and early twenty-first centuries. For now, though, we need to turn from the origin of art and job specialization to the curious nature of modern-day stress and its impact on our hunter-gatherer psyches.

## THE DRONE OF MODERN LIFE

Cars rush by outside your window, a horn blaring occasionally. The refrigerator hums in the corner of the kitchen, and the heat coming out of a duct over your head whooshes softly. Bills sit stacked on the counter, insistently waiting to be opened. A television—perhaps one of several in the house—blares advertisements from the next room, and Internet pop-up ads interrupt your attempts to check on your retirement investments. The cacophony reaches a crescendo when your spouse's cell phone rings, vibrating along the tabletop like some sort of angry digital dervish. The blare of the outside world goes on all around us, even while we attempt to focus on our "real" lives.

We are constantly surrounded by surreptitious stimuli—so much so that we take it all for granted. We are used to the notion that advertisements saturate our lives—exposure estimates for the average American range from several hundred to several thousand every day—as promoters try to sell us everything from life insurance to an enhanced sex life. Data flows at us from every direction. Information is ubiquitous and, with the rise of the Internet and broadband connectivity, more easily accessible than ever. But even things we might not think of as intrusive bombard our subconsciousnesses with stimuli. Inadvertently, the machines we have created to improve our lives may actually be causing some degree of psychological harm.

The journalist Toby Lester, in an article published in the *Atlantic* magazine in 1997, pointed out that ours is the first generation to live in an environment where background sounds from machines saturate our lives. He explained the various tones he hears at work, made up of noise from the heater, computer fan, and telephone: "My office plays a curious combination of intervals, one joyous and stable (*do-mi*), another devilish and inimical (*do-fa*-sharp), and the third (*mi-fa*-sharp) emotionally neutral. The overall result is an ambiguous chord that, at its upper end, begs for resolution."

According to musical theorists, who have assigned moods to certain chords, this combination of notes is particularly dissonant. The medieval Catholic church called the chord generated by Lester's office machines the *diabolus in musica* (devil in music), and it creates a strong feeling of unease in listeners. As Lester asks later in the piece, "Could this ambiguity and tension be one reason I so often feel on edge?" Perhaps—and it certainly provides an impetus to cherish the silence that is so rare in modern life.

This same sort of constant stimulation exists in our other sensory realms as well. We are all bombarded visually on a daily basis, and some of us experience regular assaults on our senses of smell and touch, as well (think of a crowded subway car on a hot summer day). Our lives are now lived in a state that could be called "stream of subconsciousness," as we subliminally lurch from one unrelated (and usually unwanted) stimulus to the next like floating dust particles buffeted by the random forces of air currents. Some people seem to thrive on constant overstimulation, whether dissonant or not, but most of us react rather badly to it. The end result is chronic stress of the kind that drove Michael Douglas's character William over the edge in the 1993 film *Falling Down*. Endless Los

Angeles traffic, feelings of inadequacy and powerlessness brought on by divorce and job loss, the odd combination of isolation and crowding that characterizes much of modern life—all combine to drive this character into a psychotic state, and he rampages around the city in a violent attempt to right the wrongs he perceives.

While most of us will never go this far in our reaction to the stresses of modern life, internally we are actually fighting the same sort of battle. A large body of scientific evidence shows that chronic stress damages everything from our psychological health to our sexual performance to our immune systems. Yes, long-term stress diminishes our sensitivity to glucocorticoid hormones, which are involved in inflammatory responses. In essence, our immune systems become overstimulated, fooled into thinking that there is a chronic infection or some other assault on our bodies, and the normal process by which we recognize the end of an infection is damaged. The effect increases not only our chances of catching a cold but also the likelihood of clogged arteries and autoimmune diseases.

Surprisingly, though, short-term stress is actually good for the immune system. Studies of skydivers, for instance, have revealed that the increase in their adrenaline levels adds to the number of a certain type of immune system component known as a natural killer cell. These cells are typically in the first line of defense against infections, before the rest of the immune system (antibodies and so on) has had a chance to kick into gear. Such a response would have been highly adaptive to our hunter-gatherer ancestors, since their regular exposure to short jolts of adrenaline as a result of the fight-or-flight response would have been priming their immune systems for action.

The impact of stress caused by overcrowding and geographic confinement—typically something that occurred only after the transition to agriculture—can actually be seen in some late Paleolithic sites. In these locations the population was forced to make use of a limited resource and lacked the typical hunter response to such a situation: to leave. One of these unusual sites was found along the Nile River, as described by Clive Gamble in his excellent survey of the Paleolithic, *Timewalkers:*

> The Nile (upper Paleolithic) sites show the social consequences of being tethered to such a limiting resource. Intensification in subsistence can pose other problems for survival, especially when it involves living together for longer periods. Hunters and gatherers usually cope with conflict and disagreements by walking away from the problem. However, this is not always possible. Wendorff and Close have uncovered a good deal of evidence for violence along the Nile which is not found elsewhere at 18,000 BP [years ago] in the more fertile North African refuges of the Maghreb and Cyrenaica . . . in the Jebel Sahaba graveyard at least 40 percent, regardless of age and sex, had met a violent death.

The incredibly rich Nile environment allowed the population to grow to a large enough number that such strife was inevitable. It is mirrored at other Upper Paleolithic sites with abundant aquatic resources, such as Japan and the Pacific Coast of northwestern North America. Here, as we saw in Chapter 2, large group sizes led to social hierarchies, which eventually led to fighting over limited resources. Again, it is only when you have something to fight over, and no recourse *but* to fight, that warfare becomes common. It simply doesn't make evolutionary sense to waste resources in frequent or protracted battles unless you have no other choice.

The reasons for this stressed response to living in large groups are complicated, and may have something to do with an intriguing pattern discovered by the evolutionary psychologist Robin Dunbar. Dunbar's analysis showed that the average group size among apes and Old World monkeys (the species closest to us evolutionarily) is related to the size of their brains. The larger the brain, the larger the group size, presumably because the increased neural connectivity in larger brains allows individuals to keep track of more social connections. Most species in his analysis had average group sizes of between 5 and 50 individuals. When he extrapolated the resulting line out to a human-sized brain, he found that we would be pre-

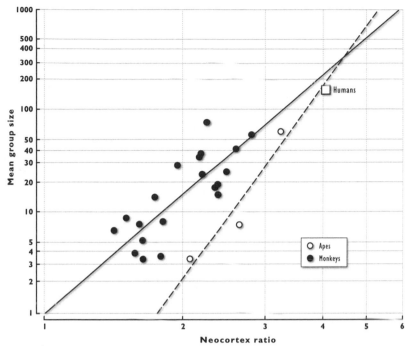

**FIGURE 24: RELATIONSHIP BETWEEN THE NEOCORTEX RATIO AND GROUP SIZE IN MONKEYS AND APES, AS DE-SCRIBED BY ROBIN DUNBAR. HUMANS HAVE A PREDICTED GROUP SIZE OF 148.**

dicted to have group sizes on the order of 150. This number turns out to be remarkably close to all sorts of natural groups that humans belong to, ranging from companies in the military to traditional Hutterite farming communities in Canada. It also turns out to be the average size of hunter-gatherer bands.

Dunbar explains that the reason humans are ideally suited to groups of this size isn't that we can't *remember* more people than this—that number runs to around 2,000—but that this is the maximum number of *meaningful* social relationships a person can keep track of. It turns out that 150 is the average number of people from whom we would be comfortable asking a favor, as well as the average number of friends and family members to whom British people sent Christmas cards in the 1990s (before the widespread use of email). Dunbar's result implies that our social structure is hardwired into our biology. If this number of people is exceeded in a hunter-gatherer band, two things typically happen. Either the group splits or, if that is impossible, it has to find some way of maintaining order by instituting the types of social structures we have in the modern world—governments, religions, laws, police forces, and so on. While the first option was open to our Paleolithic ancestors, the second is our only recourse in the post-Neolithic world.

Despite the meliorating influence of governments—what Thomas Hobbes called the Leviathan, the threatening state entity that he believed kept our "animal" impulses in check in a civilized society—there is still some psychological baggage associated with living in groups with far more than 150 people. First, it is no longer possible to treat other group members in the way we would treat them in a smaller group. We begin to dehumanize one another, and our behavior becomes decidedly unnatural. Think of standing in an elevator with strangers, where everyone tries to avoid eye contact and seems unnaturally interested in the floor number or the message they just received on their BlackBerry. No hunter-gatherer would think of not talking to other members of the group that they encountered in such a closed social situation, yet a typical city dweller often pretends that other people don't exist; to do otherwise would be overwhelming. Imagine engaging socially with every person you were in close proximity to on an average urban workday—the mind boggles at the number of possible interactions. Our brains simply wouldn't be able to handle it (not to mention our schedules), so we have developed the coping mechanism of acting as if those people aren't there, and they become yet another part of the background noise of modern life.

This psychological mismatch between the densely populated, noisy agricultural world and the sparsely populated hunter-gatherer one is almost certainly one of the reasons for the psychological unease felt by many people. Along with the other "noisy" aspects of modern life, such excessive background social stimulation is very likely part of the reason why we see increasing levels of mental illness in most societies. According to the report mentioned at the end of the last chapter, the WHO expects that by 2020 mental illness will be the second most important cause of disability and mortality worldwide. Surveys conducted in Europe and the United States show that more than a quarter of the population in any given year has had symptoms that would be diagnosed as mental illness (most are never diagnosed, however), with the most common being anxiety disorder.

This trend is reflected in the increasing use of prescription psychoactive drugs. People have always enjoyed altering their states of consciousness, with the help of substances ranging from alcohol to cannabis to the potent psychedelic *ayahuasca* used in the Amazon basin, but this is the first time in history that we are routinely drugging ourselves in order to appear *normal*. According to the Centers for Disease Control, antidepressants such as Prozac and Paxil are now the most prescribed drugs in the United States—more so than drugs for hypertension, high cholesterol, or headaches. The stimulant Ritalin, used to control attention deficit/hyperactivity disorder, is part of the daily routine of around 10 percent of American boys. Overprescription by doctors is certainly contributing to this widespread medicating, but there is clearly an underlying problem that is making us feel psychologically unwell. I would argue that this, too, is part of the continuing fallout from the Neolithic population explosion.

## INTO THE FUTURE

As we hurtle into the twenty-first century with our Neolithic baggage in tow, we are clearly still adapting to this new culture, which dates back only 10,000 years or so. The twin burdens of disease and unease that we accept as part of modern life will certainly produce profound changes in our medical systems as chronic disorders and mental illness become more and more common. New drugs will be discovered and prescribed, and we will become ever more used to living pharmacologically enhanced lives. Will we be able to find a drug to treat everything that ails us? Probably not—but the pharmaceutical companies will definitely keep trying.

There is another intriguing possibility, though, one that has become imaginable only in the past decade. This audacious new technology offers the hope of not simply treating a disease, reacting to
its symptoms, and trying, in an often hit-or-miss fashion, to cure
it. Rather, we are told, disease itself may become a thing of the
past—we will be able to anticipate rather than simply react, prevent rather than treat, and when treatments are necessary, we will
be able to target them with unprecedented specificity. This technology also, ominously, offers the possibility of eradicating diseases
forever, both in ourselves and in future generations. It is perhaps
the most potent force ever to be harnessed in the name of medicine,
and it offers us a chance to, once and for all, mold our biology to
suit the new culture we have created—to remake ourselves in the
image of post-Neolithic society. It is the field of genomics, and the
brave new world it promises is where we're headed next.

# Chapter Five
# Fast-Forward

{Eugenics} must be introduced into the national conscience, like a new religion. . . . What Nature does blindly, slowly and ruthlessly, man must do providently, quickly and kindly.

—FRANCIS GALTON,
*Essays in Eugenics,* 1909

## DERBYSHIRE

I was still groggy from my overnight flight, and as I piloted my rental car onto the M1, outside London, I had to keep reminding myself to drive on the left. I merged into the morning traffic headed toward "The North" (as the signs proclaimed in their very British way), pushed the small car up to seventy-five miles per hour, found a radio station playing classic rock hits from the 1960s and '70s, and settled in for a two-hour drive. The trip through the English countryside (albeit seen in small glimpses from the motorway) gave me a chance to ponder what I had come here to do. My other travels in researching this book had been motivated by some sort of academic interest, and in that sense this trip was different. Rather than being part of an effort to understand the scientific or technical methods behind the forces I have been discussing, this journey was much more personal. I had come here to talk to a family that had inadvertently found itself at the cutting edge of modern technology's remodeling of humanity. And my research would occur in a most unlikely place—around the kitchen table in Michelle and Jayson Whitaker's cozy nineteenth-century Derbyshire cottage.

The Whitakers were thrust into the international spotlight in 2002, when the United Kingdom's Human Fertilisation and Embryology Authority (HFEA) turned down their request to create a "designer baby," as the child was dubbed by the press. The Whitakers' first child, Charlie, had been born three years before. Although he seemed reasonably healthy at birth, by the age of three months it was very clear that something was wrong. Charlie was diagnosed with an extremely rare genetic disease known as Diamond-Blackfan anemia (DBA)—only a few dozen other cases had been found in the entire country—which meant that his blood was unable to carry enough oxygen to allow him to grow normally. The only known treatment was regular blood transfusions, but unfortunately the increase in red-blood-cell death would place enormous stresses on his liver and kidneys. Although they started these treatments immediately, and the treatments improved Charlie's symptoms, the Whitakers knew that eventually, like all other Diamond-Blackfan patients, Charlie would probably die unless a bone marrow donor could be found.

The idea was to destroy Charlie's own faulty bone marrow through chemotherapy, rendering him incapable of producing blood cells. Although this was incredibly dangerous—it's like acquiring AIDS by another means and leaves the body tremendously vulnerable to infection—it was necessary for the next step, the transplant. If a donor could be found whose immune system antigens matched Charlie's, then a sample of that person's bone marrow could be injected into Charlie's body. Hopefully, the procedure would allow these healthy cells to take up residence in Charlie's bones, eventually creating a new immune system—and healthy blood cells—which would cure Charlie's disease. However, unlike leukemia patients, who can search through the registry of millions of potential marrow donors for a match, Charlie was left out in the cold.

Even with a perfect match between donor and recipient, the chance of a successful transplant between two unrelated people is only about 15 percent, meaning that 85 percent of the time the transplant will fail. Such a risk is worth taking, however, if the patient has terminal leukemia—even a 15 percent chance is better than dying. In the case of a nonterminal disease such as DBA, though, the marrow registries don't consider the chances of success worth the risk involved in the procedure, and such transplants are not allowed. If, however, a matching sibling is the donor, the likelihood of success increases to a much more favorable 85 percent. The problem, of course, is finding a donor in the same family. Since Charlie was Michelle and Jayson's only child, their sole hope was to have another baby that could serve as a donor.

Here we enter a new realm in our journey through the fallout from agriculture and the industrialization that followed: the conflict between what we *can* do scientifically and what we *should* do morally. While the first is fairly easy to evaluate objectively—either a new technology performs its desired function or it doesn't—the second is much less clear. The divide between these two realms, the practical and the moral, will be examined more fully in the final chapter of this book. In the case of Charlie Whitaker, though, we see a clear example of how rapidly advancing technology may be outstripping society's ability to cope morally.

If the Whitakers had been faced with their decision a generation earlier, there would have been little choice. They could have had another child, hoping that he or she would have been a perfect match for Charlie. Given the way chromosomes are passed on during reproduction, this child would have had only a 25 percent chance of matching. Each additional child would have raised the odds of a match, but unless the Whitakers got lucky, they would have had to have at least three additional children to increase the likelihood above 50 percent. They had always wanted Charlie to have brothers and sisters, but they knew that each new child also carried an unknown risk of being born with DBA. Was it worth the risk of bringing another baby with a devastating disease into the world in order to treat Charlie? This question, seemingly a moral choice based on many unknowns, had a new answer in 2002 as a result of extraordinary advances in technology over the previous twenty-five years. There was, it turns out, a way to slant the odds in the Whitakers' favor.

In July 1978 a baby girl was born in Manchester, England, to Lesley and John Brown. She was delivered by Cesarean section and weighed five pounds, twelve ounces at birth. Nothing particularly remarkable—another in the long line of children born that summer in England, the last before the "Winter of Discontent," which would usher in the Thatcher government the following year. But the baby, named Louise Joy, would herald the dawn of a new era in human reproduction: Louise was the first child ever conceived by in vitro fertilization (IVF)—the original "test-tube baby."

The doctors who performed the IVF to create Louise had been inspired to try the technique in humans by its tremendous success in other animals, particularly rabbits. While the laboratory rabbits didn't really need any help having babies, many people did, including the Browns, who had been trying to conceive for nine years. The success of their pregnancy with Louise led Lesley and John to have another IVF baby a few years later, and the new world was off and running.

The wonderful thing about IVF is that it simply helps nature do what isn't happening naturally. No modifications take place, and the natural sperm and egg of the parents are used in the process. Lesley, it turned out, produced perfectly normal eggs, but her fallopian tubes were blocked, preventing the eggs from encountering the sperm. By bypassing this anatomical impasse, IVF allowed Lesley to do what her body hadn't—conceive a child with her husband. No other modifications of the normal process of human reproduction took place, and the implanted egg was every bit as much of an unknown as a child conceived by more typical methods.

Since the late 1960s, scientists had been making use of a little-known fact about animal embryos to develop a much more radical approach to IVF. Adults, children, and advanced embryos are incapable of regenerating large quantities of lost tissue—you can't regrow a new arm if yours is severed—but embryos at the very earliest stages of development seem to have no problem doing so. Once fertilization takes place and the fertilized egg starts to divide, there is a doubling in cell number about every eighteen hours, so that on the morning of day three after fertilization there are eight cells in the embryo (2 x 2 x 2). If a cell is removed from the embryo at this point, it seems to make no difference to the further development of the embryo—the remaining seven cells continue to divide, forming the complex structures of the advanced embryo and, ultimately, the recognizable tissues of the fetus. The cell that is removed, though, contains the complete genome of the developing embryo. With advances in molecular biology in the 1980s, particularly the development of the polymerase chain reaction, which allows small quantities of DNA from a single cell to be studied, scientists were suddenly able to read the genetic information in this cell, opening up an entirely new era of childbirth.

This technology, known as preimplantation genetic diagnosis (PGD), allows the IVF team to predict the characteristics of the future child. In the case of the Whitakers, by creating a large number of eight-cell embryos and testing all of them for the genes that would determine whether the child could serve as a match for Charlie, they could bypass the laws of probability and implant only those embryos that could serve as donors. Such a technique is not without its detractors, however, which is why the process needed to be approved by the HFEA. In the United Kingdom, unlike in the United States, all of these new technologies are strictly regulated to protect the rights of the unborn child. When the Whitakers' application became known, many people came out in opposition. Wary of letting anyone create an "organ donor" baby, some opposed the procedure on ethical grounds. They assumed that the Whitakers were simply conceiving the child to serve as Charlie's tissue donor, an idea the press had a field day with. The HFEA had been burned by a similar case the year before, in which it had authorized another U.K. family to use PGD to create a donor for their child and had subsequently been sued by a group calling itself the Comment on Reproductive Ethics. In part as a reaction to this, the HFEA refused to give the Whitakers permission. The new technology might have been possible, but it wasn't acceptable, at least to many in the United Kingdom. The *Daily Mail,* a national newspaper, reported the decision on its front page with the headline "THE LITTLE BOY THAT SCIENCE WON'T HELP."

The Whitakers' physician, Mohamed Taranissi, was one of the

U.K.'s most successful fertility doctors. He had been hoping to use Charlie's case to raise awareness about the availability of such technologies, and eventually to implement them on a larger scale throughout the country. In 2002, however, many in the medical community still considered PGD very much an experimental procedure. For this reason Mr. Taranissi (British surgeons are referred to as Mr., rather than Dr.) had planned to fly in a team from a private fertility clinic in Chicago to perform the procedure in the United Kingdom. But when the HFEA failed to approve the procedure, he was faced with a dilemma. So, in the best tradition of the American can-do attitude (despite being English), he suggested to the Whitakers that they simply get on a plane, fly to Chicago, and do the whole procedure there.

Jayson and Michelle Whitaker are not rich, but when they made the decision to undertake PGD—a very expensive procedure that can cost over $30,000 per attempt, often with several attempts needed to produce a viable pregnancy—they were willing to mortgage their house and borrow money from family members. Mr. Taranissi, though, made them an offer they couldn't refuse: he would pay for all of the expenses incurred, including the trip to Chicago. So, three weeks after hearing that their HFEA application had been rejected, they found themselves undertaking last-minute hormone tests at six A.M. in London before rushing out to Heathrow to board a plane to Chicago. The clock was ticking.

They arrived on a Saturday, and early on Monday morning Michelle's eggs were harvested—thirteen of them. Five of these died over the next few days, leaving eight that could be fertilized. After the specialists mixed them with Jayson's sperm, the Whitakers sat back and waited, fingers crossed, hoping for the best. On day three, six of the eight embryos were still viable, and the IVF team carefully cut one cell away from each. The genetic tests revealed that three of them were a match for Charlie—a stroke of good luck, since the chance of a match had only been 25 percent. Michelle and Jayson decided to implant two, in case one didn't take. With the deed done, they boarded the flight back to London. On the trip home, Michelle, feeling the telltale signs of morning sickness—perhaps psychosomatic, perhaps real—knew that she was pregnant. When they arrived in the United Kingdom, they immediately went to see Mr. Taranissi, and the tests of her hormone levels suggested that only one embryo had implanted. As he told them, given her earlier success at getting pregnant naturally and her relatively young age (she was thirty at the time), "if this doesn't work for you, it won't work for anyone." It did.

The pregnancy was soon confirmed by ultrasound, and the Whitakers were on their way down the path toward Charlie's treatment. At eighteen weeks they had amniocentesis performed to confirm the match but chose to learn nothing else—not even the baby's sex. With this final confirmation, all they had left to do was wait for the child to be born. It was a normal pregnancy, the only complication being that the child didn't turn around in the uterus; it was set to be a breech birth. For this reason the Whitakers' obstetrician decided to perform a Cesarean delivery on June 16, 2003.

Apart from delivering a healthy baby, the goal of the medical team—and of the whole procedure—was to harvest some of the child's cord blood. The umbilical cord, which serves as a conduit for oxygen and nutrients from mother to child, also contains a particularly rich source of hematopoietic stem cells. These cells are unique in that they can differentiate into the bone marrow cells that in turn produce normal blood cells. In effect, they serve as a proxy for a bone marrow transplant, and using them involves none of the invasive techniques used to harvest bone marrow (painful needle punches through the sternum being the normal means of bone marrow collection). After the doctor introduces them into the recipient in the same way as an ordinary blood transfusion, they manage to migrate to the bone marrow and proceed to differentiate into normal bone marrow cells. All that is required is the withdrawal of blood samples from the umbilical cord after the birth, once the cord itself has been cut. It is minimally invasive, and is now done at many births, so that cord blood can be saved for future therapeutic use. Jayson himself had been trained to collect the cord blood in case the delivery happened outside the hospital; it was simply too valuable to risk losing.

The Cesarean delivery went smoothly, and around 150 milliliters of cord blood—"half a can of Coke," as Jayson describes it—was collected. This was immediately washed and the cells were frozen, as it would take at least six months to confirm that the baby boy, named Jamie, didn't also have DBA. Luckily, he didn't, and during the winter of 2003–04 the Whitakers and Mr. Taranissi began planning Charlie's transplant.

They settled on the summer, since it was the "low season" for colds and flu, and because what would come next would be the most dangerous part of the whole procedure. Using chemotherapy, they would kill all of Charlie's existing bone marrow cells. This was a precondition for the transplant to work; it would also, however, render him susceptible to opportunistic infections. In order to minimize the risk, he would have to be kept in isolation while it was being carried out. As Jayson described it to Charlie when they explained what was coming, they had to "put his soldiers to sleep so they wouldn't attack Jamie's soldiers."

Charlie entered the hospital in late June, and the first drugs were administered soon afterward. Michelle described to me in detail how heartbreaking it was to see her five-year-old son screaming from the pain of the chemotherapy, losing weight and, eventually, his hair as he became weaker and weaker. Finally, after ten horrible days, Charlie's neutrophil count had dropped to zero—he no longer had a functioning immune system. (Neutrophils are a particular type of white blood cell involved in the immune reponse.) At this point they could stop the chemotherapy and perform the transplant.

After all of the steps leading up to this point, the transplant itself was positively mundane: the cord blood was simply thawed out and transferred into Charlie's body through a vein in his arm. The most unusual thing about the procedure was that it left Charlie smelling like canned corn, a result of the substance used to freeze the cells. Half an hour after it started, the procedure was over—all that was left was to wait and hope.

The cells in the transfusion would, in theory, find their way to Charlie's bone marrow, implant themselves there, and start making blood cells again. They wouldn't produce Charlie's DBA cells, though—they would make healthy blood cells like Jamie's. If it was going to work, it would happen in a matter of weeks. And it did—before the end of July, the first neutrophil appeared in Charlie's blood tests. Hoping to reduce his likelihood of catching a hospital infection, Michelle and Jayson took Charlie home on July 30, just over a month after he had entered the hospital.

Over the next few months Charlie was slowly weaned off the drugs used to prevent rejection until, at six months after the transplant, he was off all of the medications. The ordeal of the transplant was over, but it still wasn't certain that he had been cured. Bone marrow biopsies were performed at six months and one year to confirm that Jamie's cells living in Charlie's marrow were healthy, and the family waited to see if any of the symptoms of DBA would return. Luckily, they didn't—Charlie's hematocrit level, a measure of red blood cell production, remained within the normal range. Finally, in March 2007, Charlie was declared officially cured. The long ordeal had most definitely been worth it.

As the Whitakers were nearing the end of recounting this amazing story to me, Jayson started to talk about the difficulties they had encountered from the HFEA and other groups overseeing the ethics of PGD. As he saw it, if there was a medical solution available to a problem, you should be allowed to take advantage of it. "We're not playing God," he noted. "All we're doing is changing the odds." He had since been appointed a member of the Human Genetics Commission, set up by the U.K. government to study the risks and benefits of new genetic technologies. When I asked him how he felt about choosing other genes using PGD, he told me he was hesitant to do it for something like hair color, but that it should be allowed for genes that might affect the risk of diabetes, for instance. If the information is available, he said, people will want to have access to it.

As I was leaving the Whitakers' house I snapped a quick picture of Charlie and his brother, who were playing in the front yard.

FIGURE 25: JAMIE AND CHARLIE WHITAKER.

They seemed so completely normal that I found it hard to believe what I had just heard, a tale that mixed cutting-edge science, medical ethics, and the media against the backdrop of a very sick little boy. I'd had a glimpse of a future that will become more and more common as these new reproductive technologies and genetic testing become more widespread. Charlie and Jamie seem unaffected, but the question remains: Will these medical advances end up changing us in ways that we can't anticipate?

## AN ACCELERATING TREND

As I mentioned above, the first application of IVF to human preg nancy was in 1977. Since then the use of the technique has exploded, aided by two trends. The first is greatly improved methodology that has seen the likelihood of a successful IVF pregnancy increase dramatically since the early 1980s, so that today the chances of success—a live birth—approach 40 percent per treatment cycle in women under 35 and can be as high as 10 percent in women over 40. The second is an increase in the age at which women first get pregnant. In the United States in 1971, the average age at which a woman first became pregnant was 21.4 years, while in 2003 it was 25.2. In Britain, the figure in 2003 was 27.4 years, and in Switzerland, 28.7. This trend is even more pronounced in women with a college education. Between 1960 and 2003, the percentage of first births for women over 30 tripled from 7 percent to 22 percent, and among women with higher educational attainment (college and graduate school) it is becoming more and more common to delay the decision to have a child until after the age of 35.

Of course, fertility drops significantly after age 35, making conception less and less likely, and this explains why the use of IVF has increased so dramatically over the past twenty years. It's estimated that as many as 4 percent of births in some European countries are IVF-assisted. More than one million babies worldwide have now been born using the procedure, and the trend is accelerating. Today, 10 percent of American women over the age of 35 and 22 percent of women over the age of 40 use IVF techniques to conceive.

The cost of in vitro fertilization is typically $10,000 to $12,000 per cycle, and with increasing age and reduced likelihood of pregnancy, more cycles are required to guarantee success. This clearly isn't something to undertake on a whim. With this in mind, many more couples are now opting for PGD in addition to IVF, hoping to increase their chances of a successful transplant. While a 2004 study in the journal *Human Reproduction* showed that PGD ap-

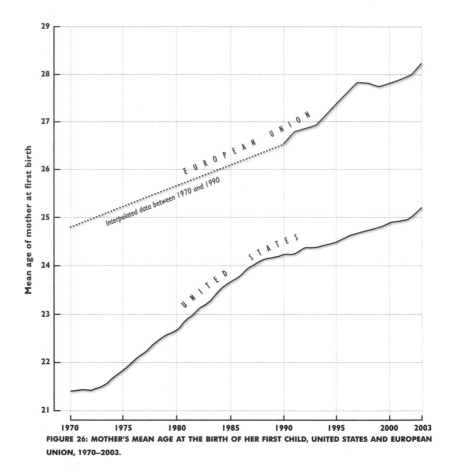

**FIGURE 26: MOTHER'S MEAN AGE AT THE BIRTH OF HER FIRST CHILD, UNITED STATES AND EUROPEAN UNION, 1970–2003.**

peared to have no effect on the success rates of IVF in 37-year-old women, the procedure is still increasing at a rate of 15 to 30 percent per year, according to the *Los Angeles Times*. Clearly, the technique will soon be far more widespread, as birth trends project ever-growing numbers of older women conceiving via assisted reproductive technologies.

Typically, unless the parents know about a severe genetic disorder like Tay-Sachs or cystic fibrosis, all that is tested in PGD is something known as *aneuploidy*—an abnormal chromosome number, such as that responsible for Down's syndrome. However, as the Whitakers' case showed, it's quite possible to test for other genetic traits as well. And as we enter the new era of genomic knowledge that is set to unfold over the next decade, more and more will be understood about how our genes influence a variety of traits, not just relatively rare genetic disorders. With the increasing reliance on IVF, and its expense, it seems likely that more and more couples will opt for the perceived benefits of genetic enhancement.

Imagine yourself in the position of a couple in which one member is a well-educated woman in her mid-thirties who has delayed having children in order to pursue her career and now earns a comfortable salary; the husband has a similar type of job. They own a house in a safe suburb with good schools and are health-conscious—they are exactly the sort of parents that most state child welfare administrators would wish on a child. The couple will likely not have more than two children, and may opt for a singleton in order to devote their time and resources to giving the child every advantage they can. They may employ a well-qualified nanny to look after the child if the mother returns to work soon after the birth, and will certainly seek out good preschools and kindergartens. As the child gets older they will find tutors if the child needs them—and often even if he or she doesn't—and enroll him or her in summer enrichment courses when school isn't in session. Their thoughts will also turn to sports successes, or perhaps music lessons, always with the goal (either stated or tacit) that the child should be encouraged to succeed to the best of his or her abilities. Ultimately, college preparation will rear its head, in some cases as early as primary school, as the parents try to anticipate which secondary school will best prepare their offspring for a good university. Once a suitable institution is found, the parents may feel that they have fulfilled their role of sending their child on the way to a happy and successful life.

Even if this journey unfolds smoothly, it will cost hundreds of thousands of dollars over the child's lifetime. Add to that the cost of conception (if assisted) and pregnancy, and it's understandable if some parents start to see the whole process as an investment—a worthwhile one, to be sure, but a very real expenditure of scarce resources for the next generation. There is also, perhaps, a certain hubris—a desire to see one's own offspring reflect well upon oneself—which, compounded with the other factors, can produce the complex path to adulthood taken by many upper-middle-class children today.

Along this well-trodden path, if any sign appears that the desired result might not come to pass, the parents may take action. A young boy's rambunctious personality may be calmed by daily doses of Ritalin, a little girl's crooked teeth corrected with braces, and any sign of a serious medical condition attacked with the latest high-tech diagnostic tools and a dedicated medical team. We all know that life's uncertainties can blindside even those with the best-laid plans. But what if there was a way to avoid this, and stack the deck in your favor, by choosing the genes in your offspring? While some people might find this unethical, as we saw with the Whitakers, others simply see it as an intelligent use of modern technology.

This is the future promised by new genetic technologies. Although we know a fair amount about the genetic basis for many rare disorders—those that occur in less than 1 percent of the population, like Tay-Sachs or cystic fibrosis—geneticists are still very much in the process of discovering the genetic factors that influence the common diseases we learned about in Chapter 3, such as hypertension and diabetes. The environment might play the most important role in determining whether you get these diseases, but there are certainly genetic factors as well. And for psychological traits like schizophrenia, bipolar disorder, and alcoholism, even less is known. We are currently witnessing a revolution in our understanding of the genetic bases for such illnesses, and we will be discovering many of them in the next decade. In an interview in *The Scientist,* Eric Lander, one of the leaders of the Human Genome Project and the director of the Broad Institute in Cambridge, Massachusetts, has described this period in the field of genetics as "a tremendously exciting time to study human variation, because we finally have enough tools and infrastructure to correlate genotype to phenotype" (phenotype is the ultimate effect of a gene—its influence on appearance, disease state, behavior, and other outwardly visible characteristics). Clearly, we are on the cusp of a revolution in our understanding of our own genetic predispositions, and once we have the information, many people will want to use it, in the same way that other technologies have been embraced. But what will this information look like, and how will we make these decisions?

## TWENTY-FIRST-CENTURY NEEDLES

There are around 23,000 genes in the human genome. This figure still seems startlingly low to me, since throughout my career as a graduate student and postdoctoral research fellow we had always estimated the number to be close to 100,000. When the draft of the human genome was announced in 2000, though, it was clear that we are far less complex at the genetic level than we'd thought. Or are we?

What the gene number revealed (incidentally, fruit flies have nearly as many genes in a genome that is one-twentieth the size of our own) was that it wasn't simply large *numbers* of genes that produced human complexity; what was significant was the *way* the genes were being turned on and off to create our cells, tissues, and other characteristics. As the saying goes, it's not about the size, it's how you use it. Clearly, the human brain is far more complex than that of an insect; the difference lies not in having lots of brain-specific genes (memory genes, speech genes, genes that cause us to appreciate jazz, and so on) but, rather, in using the genes we have in exceedingly complex ways to produce these traits. The complexity of these gene interactions reveals an insight into an important genetic phenomenon known as *pleiotropy*.

Pleiotropy refers to the many effects of a gene on the organism's phenotype, apart from the ones that are expected from the actual function of the gene. For instance, a single letter change in the sequence of the hemoglobin gene that codes for the oxygen-carrying protein in our red blood cells can cause the symptoms that we know as sickle-cell anemia, which occurs at a frequency of around 8 percent in people of African descent. These symptoms include kidney failure, stroke, and liver damage. All are the result of a single nucleotide change that causes the hemoglobin molecule, under certain circumstances, to change its shape and form crystalline structures in the blood cells, rendering them rigid and unable to carry blood into the smallest capillaries of the body. A seemingly small change, but one with significant effects.

Interestingly, this same change also renders people who carry the mutation less susceptible to infection by the malaria-causing parasite, in much the same way as the *G6PD* mutations we learned about in Chapter 3. In one of those odd twists of evolutionary fate, sickle-cell carriers—people with only one mutated copy of the gene, rather than the two necessary for full-blown anemia—are actually at an advantage in the malarial jungles of central Africa. This accounts for the relatively high frequency (around 25 percent) of carriers in this population, as thousands of years of malaria exposure have selected for this variant in the people living there. Something that is bad in one context turns out to be good in another.

Similarly, trisomy 21—the presence of three copies of chromosome 21—causes the congenital disorder known as Down's syndrome. As far as we can tell, the genes present in the extra copy are identical to those present in one of the other two—it's a duplicate of one of them. Why this duplication causes the suite of symptoms (lowered intelligence, motor dysfunction, facial abnormalities) seen in Down's cases is unknown, but it almost certainly has something to do with the number of copies of certain key genes, which affects the way in which they function. These genes, located in the so-called Down's critical region, are sufficient to cause the disorder even if they are the only parts of the chromosome that are duplicated, copied, and inserted into the otherwise normal chromosome. Why these genes are so finely tuned as to require exactly two copies, but not three, in order to function properly is currently unknown; clearly, seemingly minor genetic changes can have large unintended effects. In this case it is probably a change in gene regulation; the level at which one (or more) of the genes in this region produces its encoded proteins is too high with three copies, and developmental problems ensue.

Disorders such as sickle-cell anemia and Down's syndrome, despite the complexity of the functional differences by which small genetic lesions are converted into the final phenotype, are still relatively simple by the standards of most diseases affecting humans. The big killers we learned about in Chapter 3 also have genetic components, but instead of a single gene or chromosomal region, they involve small changes all over the genome, like tiny needles hidden in a gigantic haystack. Heart disease, for instance, has had dozens of genetic mutations associated with it in the scientific literature over the past twenty years. Some of these have stood up to scrutiny, while others have not been confirmed by other research teams and remain uncertain. And even for the confirmed associations, the relative odds are typically quite low, increasing the risk of hypertension or a heart attack by 50 percent or so. In other words, if a person not carrying the associated genetic variant has a risk of, say, 10 percent of having a heart attack at some point in his or her life, then someone carrying the associated mutation has a 15 percent chance. This means that there is still an 85 percent chance that he or she won't have a heart attack, making the interpretation of the genetic effects quite difficult.

The environment is another complicating factor. As we saw with the Pima Indians, people who are very similar genetically can have vastly different disease rates, depending on their lifestyle. Even if you do have genes that predispose you to get diabetes, for example, living a healthy, active lifestyle low in high-calorie foods like sugar and fat will make you much less likely to develop the disease. These environmental effects further complicate the analysis of so-called complex, or mutifactorial, diseases like hypertension, diabetes, stroke, and the other big killers in the developed world. It is incredibly difficult to tease apart genetic factors from shared environmental factors. Suppose, for instance, that you are a researcher who has noticed a high incidence of heart disease in people of Scandinavian descent living in one part of Nebraska. Is this because of shared genetic factors—after all, they all come from the same part of Europe—or shared environmental factors, such as diet and lifestyle? The genetic studies need to be designed very carefully in order to separate these various influences, and then large numbers of people need to be analyzed in order to find a genetic association that is statistically significant.

The end result is that genes alone don't tell the whole story. Ultimately, your DNA influences your health, but it doesn't determine it. The real advantage of genetic testing, at least if done early enough in life, is that it can provide an insight into some of your risk factors, in the same way that analyzing your diet and lifestyle can. Armed with more information, you can make intelligent choices about the way you live. Carrying risk factors for heart disease? Probably even more important for you to exercise regularly, eat right, and never start smoking. High risk of prostate cancer? Maybe you should start getting screened at age forty, rather than the currently recommended fifty. Low risk of diabetes? Maybe you don't have to be as careful as some other people about eating sugary foods (though you should still watch what you eat).

Such testing is becoming more commonplace as more of the genetic factors are discovered. Several companies now offer genetic testing to assess your risks of developing various disorders. All of the tests are predicated on the notion that knowledge is power, and that knowing will allow you to make lifestyle changes to reduce your odds of developing the diseases in question. If you are over the age of sixty, you already have a lifetime of behaviors that can't be undone, so the testing is less useful than it is to someone in their twenties. Behavioral recommendations have changed dramatically over the past fifty years as epidemiologists have uncovered various environmental risk factors. Smoking was thought to be only vaguely unhealthy in the 1950s, a suntan was a key component of a healthy appearance in the 1970s, margarine was touted as the healthful alternative to butter in the 1980s, and a high-carbohydrate diet was considered desirable in the 1990s. We all have made lifestyle decisions based on these recommendations, often blindly following medical advice doled out with little underlying data to support it.

Our genetic risks have been a big unknown through all of this. While family history is often taken into account when medical recommendations are made, a lot of advice is based on the assumption that everyone responds in much the same way to whatever the factor is. Clearly, though, this isn't the case—we all know people who smoke for years and never develop lung cancer, or who can eat anything they want and never become fat, or who do exactly what they should do and still die young. Jim Fixx, for instance, the man whose *Complete Book of Running* popularized jogging in the 1970s, died at fifty-two of a massive heart attack despite being extremely fit and having competed in dozens of marathons and other long-distance races. His father had died of a heart attack at forty-two, and clearly Jim was carrying genetic variants that predisposed him to heart disease despite his healthy lifestyle.

Progress in the field of genetic disease associations has been extraordinary since the completion of the Human Genome Project. In a March 2008 American Heart Association press conference to announce a number of new associations, Maren Scheuner of the RAND Corporation noted that fifteen years earlier there were only one hundred known genetic associations with disease (and many of these, such as cystic fibrosis, were for relatively rare disorders), while the number at the time of the press conference was around fifteen hundred—and growing. The rate of discovery is so high that health-care professionals simply cannot keep up, and one of the major challenges of the next phase of genetic medicine will be educating the health-care workforce.

The most useful time to test people, of course, would be at birth. This will probably start to happen in the next decade, once the efficacy of such testing becomes clear to the medical community. Lifetime risk could be calculated and a lifestyle "prescribed" that takes into account your genetic risk factors. It's also possible, though, for testing to be carried out before conception, at the embryo's eight-cell stage, using PGD. It's now clear that many diseases are influenced by the environment in the womb, and such knowledge would allow the mother to modify her pregnancy to suit her offspring's genetic makeup. For instance, Ellen Ruppel Shell, in her book *The Hungry Gene,* discusses the Dutch "hunger winter" babies. These children were born to women who were pregnant during the winter of 1944–45, when a wartime famine affected the population of Holland. Those whose mothers had lived through the famine during the first two trimesters of their pregnancy were 80 percent more likely to be obese as adults, and they also showed higher rates of diabetes and other chronic diseases. Scientists now understand that the fetal environment can affect how genes are expressed, and knowing the risks prior to becoming pregnant might help a woman tailor the womb environment to the child's genetic inheritance.

There is, however, another possibility—one depicted in the 1997 science fiction film *Gattaca,* in which most children are created via IVF and a genetic scan allows parents to choose the genes that are most desirable. While such a scenario still seems futuristic, the Whitakers have shown that in fact it is not only possible but already happening. Michelle and Jayson simply tested for genes that would determine the success of Charlie's stem cell transplant, but it's possible to test for any of the genetic variants that might contribute to disease. Why saddle your offspring with the burden of being genetically predisposed to Alzheimer's, for instance, when you can simply select for the embryos that don't carry the predisposing variant in the *ApoE* gene?

Although some people may object to such genetic selection on the grounds that it is unethical, we already allow many other diagnoses to be carried out prior to birth, in the name of prevention. This is the thinking behind amniocentesis and chorionic villus sampling in mothers over the age of thirty-five. If a disorder like Down's is seen in the fetus, the pregnancy can be terminated and the couple can try again. More than 90 percent of couples faced with such a diagnosis choose this option. With the application of PGD, of course, such invasive procedures and emotionally difficult decisions can be avoided altogether. A 2001 court case in France, where a child born with Down's syndrome successfully sued his mother's physician for allowing him to be born with the disorder, even raised the possibility that such testing could eventually be required—if not legally, then at least *de facto*, in order to avert the risk of a costly lawsuit.

With health-care costs in America currently increasing at 11 percent per year (and at least half that amount in most other developed countries), a rate far beyond the broader rate of inflation, in the future such testing may also be required by health insurance companies if you want your child to have insurance coverage. As a result of rising health insurance costs, some American employers, such as the Michigan-based health-care company Weyco, have begun firing their employees for engaging in risky behavior like smoking. Will willfully ignoring your family's genetic risks eventually become grounds for dismissal as well? Even in countries that have universal health care, there is still a strong incentive to prevent rather than treat disease, and the ultimate form of prevention is to avoid risks altogether.

Even though we can enact legislation to protect the rights of people with genetic disorders, such as the Genetic Information Nondiscrimination Act, which was passed by the U.S. government in 2008, I suspect there will eventually be social pressure to test for and act on genetic information. Such pressure—and the desire for a healthy baby, of course—is the reason most pregnant women in the developed world over the age of thirty-five choose to get their fetuses tested for chromosomal abnormalities; nearly all have ultrasounds and other tests. Will significant numbers of people choose to use PGD in the future in order to ensure that they have a normal child? It certainly seems possible, as a recent U.S. survey suggested: 52 percent of those polled said they would use a prenatal genetic test for susceptibility to heart disease, and 10 percent and 13 percent would test for genes associated with height and intelligence, respectively. A new project in China aims to identify gifted children and nurture their future development based on the results of genetic tests. As the project's director says, "Nowadays, competition in the world is about who has the most talent. We can give Chinese children an effective, scientific plan at an early age." And the California-based clinic Fertility Institutes even announced in early 2009 that it would begin offering PGD for traits such as hair and eye color. Clearly, the genetic future has arrived.

For those of you who have been following my argument in this book, that our biology is mismatched with our culture, it's clear that selecting the genetic traits in our children might be leading us further in the same direction we have been headed in for the past 10,000 years—only much more quickly. As Francis Galton, the father of eugenics, argued in his century-old quote at the head of this chapter, we are simply doing what nature (or culture) has done for many, many years. Using our knowledge of genetics, we will be able to select for traits that will further adapt us to the culture we have created. While the lactase persistence variant that Pritchard analyzed in Chapter 1 may have taken thousands of years to increase to its present high frequency in milk-drinking populations, in theory we can accomplish a similar feat in a few generations—or even a single generation, if applied universally—using new genetic diagnosis techniques. Although these decisions may seem very personal, motivated by parental love and the desire for a healthy child, in fact they have a much larger effect. This is because by selecting for certain traits we are not merely affecting our child but also all of his or her offspring. In the same way that the transgenerational power of developing agriculture during the Neolithic period set in motion forces that would play out over thousands of years, so too will choosing the genes of our children. In effect, we will become the agent of selection for future generations. Such a statement may sound extreme, but it is nonetheless true; once genetic variants are removed from a population, either by natural or artificial selection, only additional mutations can reintroduce them. And that raises the specter of where this may be leading us.

## CAREFUL WHAT YOU WISH FOR

The great jazz saxophonist Charlie Parker lived during a golden time in music history—what is commonly known as the "bebop era," lasting from 1945 to 1960. Parker and his fellow musicians created a revolutionary new form of music, characterized by lengthy improvised sections in the middle of an otherwise composed piece of music. Skillful improvisations are the hallmark of the great jazz musicians of this period, who included legends such as Miles Davis, Thelonious Monk, and John Coltrane. Parker, as with many of his fellow jazzmen, took great pride in living the life of "sex, drugs, and rock 'n' roll," as it would come to be known in the 1970s. He may have taken it to extremes, though. According to the psychologist Geoffrey Wills, in a 2003 paper published in the *British Journal of Psychiatry,* he "consumed enormous quantities of food, used heroin in increasing amounts, was known to drink 16 double whiskies in a 2-hour period and entered into hundreds of affairs with women." Art Pepper, another jazz great from the period, was quite explicit about his own appetites: "I always prided myself on being able to stay up longer than anybody else, drink more than anybody else, take more pills, shoot more stuff, or whatever."

These men were not anomalies. According to Wills, the vast majority of great jazz musicians from this period succumbed to drug addiction, alcohol abuse, schizophrenia, bipolar disorder (manic depression), or some other serious mental ailment. The most common factor in these problems, according to Wills, was disinhibition—the lack of a behavioral "edit" mechanism, by which most people limit their activity to what is generally thought to be healthy and socially acceptable. Perhaps not surprisingly, a recent study by Charles Limb of Johns Hopkins and Allen Braun of the National Institutes of Health that used magnetic resonance imaging to peer into active brains showed that the part of the brain involved in inhibition is turned off in improvising jazz musicians. Perhaps being able to turn this part of the brain off is what allowed the greats to become so good at this critical skill, and this characteristic extended into other parts of their lives as well, leading to drug abuse and promiscuity.

Similarly, many great artists and writers have been prone to alcoholism and other mental disorders. Manic depression has been linked to high levels of creativity, as detailed by Arnold Ludwig in his book *The Price of Greatness.* Overall, there seems to be a correlation between mental illness and creativity that goes beyond the merely anecdotal. It's as though either something about the act of creation predisposes a person to such problems or, perhaps, that having such a predisposition leads to greater creativity. Just like sickle-cell anemia, where a genetic variant that is bad in one context can be good in another, it's possible that the genetic variants that predispose a person to mental disorders may also foster the type of nonlinear thinking that leads to great artistic accomplishments.

The psychologist David Horrobin, in his book *The Madness of Adam and Eve,* does, in fact, argue that the genes for schizophrenia are the same as those that produce creativity. While this model has not been widely accepted, he does make a compelling case for the oddly similar incidences of schizophrenia in worldwide populations. Around 1 percent of people, regardless of ethnic background or geographic origin, are schizophrenic, whereas most other human diseases vary widely in their population incidences. He suggests that this is because schizophrenia results from having too many copies of "creativity genes" that, in smaller numbers, are useful and have been selected for in human populations because of their beneficial effects. In other words, if you have one or two copies of genetic variants that predispose you to schizophrenia, you might also be predisposed to become a great composer or mathematician (or, to put it in the context of our Paleolithic ancestors, better at developing new tools or anticipating where to find food); having three or four, however, could tip you over the edge to schizophrenia. Again, such a model is highly speculative, but it would explain both the relationship between mental illness and the creative process and why all human populations exhibit schizophrenia. Dr. Navratil, from the House of Artists, certainly would have agreed with Horrobin.

But what might happen in a future where such variants are routinely selected against? After all, most parents wouldn't wish a debilitating illness like bipolar disorder or schizophrenia on their children. In so doing, however, might they also be selecting against creativity? While such a one-to-one correspondence between genetic predisposition and ultimate outcome vastly oversimplifies the complexity of human behavior and psychiatric disorders, it is likely that extreme creativity has at least a partial genetic basis. Perhaps creativity is a knife edge, on which we sit poised to teeter in the direction of either illness or accomplishment—and, indeed, many creative people alternate between the two. In an effort to avoid the former, could we also be deterring the latter? After all, at its core, creativity is imagining things that aren't there and then making them real. Schizophrenia is largely defined by such vivid imaginings, although to such an extent that it becomes deleterious to the individual.

Psychology isn't the only thing we might be able to influence. As we saw with James Neel's "thrifty genotype" hypothesis, if we select for genetic variants that protect against diabetes, will we also be creating hothouse flowers of ourselves—svelte, calorie-burning humans with metabolisms perfectly adapted to the excessive nutritional intake and low exercise levels of modern industrial society? If so, will we lose our ability to cope with any potential disaster that may befall us in the distant future? As our food supply tends toward an ever greater monoculture of clonal genetic strains, what will happen if the equivalent of the Irish potato famine strikes in several hundred years' time? Will our distant grandchildren thank us for having molded them into the perfect twenty-first-century biology—one that may lead them to starve in a time of scarce resources?

The real threat of such genetic decisions will, of course, take many generations to be ascertained. The world may change so completely in the next few hundred years that such self-selection will become less important. Perhaps famine really will become a distant memory, and computers will be more creative than humans. But another option, one that manifests itself in genetic changes within our own lifetimes, is now being tentatively applied to people born with terminal illnesses. It is the option of direct genetic modification, and its aim is to eradicate serious illness not by selecting for genetic variants in the next generation but by substituting a good for a bad gene within a living person. Such treatments, and their potential consequences, raise many questions.

## VIRUSES, ANTS, AND REPUGNANCE

W. French Anderson, who was the scientific consultant on the movie *Gattaca,* leapt to worldwide fame in 1990 when he became the first doctor to treat a genetic disorder with gene therapy. His patient, a four-year-old girl from Cleveland, Ohio, named Ashanthi DeSilva, had a rare disorder known as severe combined immunodeficiency, which is like an inherited form of AIDS in which the person has no functioning immune system. In Ashanthi's case, it was caused by crippling mutations in both of her copies of the gene for adenosine deaminase (ADA), an enzyme critical for immune function. Anderson inserted a normal copy of the gene into a retrovirus, which was then used to infect a sample of her white blood cells. These modified white blood cells were then transfused into Ashanthi, the goal being to introduce the functioning gene into her genome and overcome the genetic deficit by overproducing a functional copy of the key gene. Although her condition improved dramatically and she's still alive as of 2009, it has been hotly debated whether the gene therapy or the additional treatment provided by administering extra ADA by injection was responsible—the ethical review board had refused to take the chance of letting the gene treatment work on its own.

The success of Ashanthi's treatment led to an explosion of applications for additional gene therapy trials. By the end of the 1990s, around three thousand people had received treatment using these methods, but, unfortunately, most weren't as lucky as Ashanthi. When, in 1999, eighteen-year-old Jesse Gelsinger died of a massive immune response to the adenovirus used to carry a therapeutic gene to his cells, it not only shut down all such work at the University of Pennsylvania but led many people to question the safety of the whole field of gene therapy. If the risks were so significant, was it even worth undergoing a therapy that had a minimal chance of succeeding?

The mixed results of gene therapy expose another case of pleiotropy. In this case, what seemed to work in theory or in the laboratory (though two monkeys given Jesse's treatment had died of similar complications) simply didn't work when used in a human being. The multiple levels at which gene therapy components interact with human physiology makes predicting outcomes very difficult. More and more, we are coming to realize that tinkering with nature can produce unintended effects, even if the tinkering seems well planned and justified.

One recent example from a different field was described in a 2008 study published in the journal *Science.* The researchers were testing different methods of preserving acacia trees on experimental plots in Kenya. Half of the plots were surrounded by fences that excluded herbivores such as elephants and giraffes, while the other half were left open. The thinking was that if herbivores were kept away, at least for a while, the acacia trees—which had become stressed by overgrazing—would have a chance to thrive. Quite the opposite happened, unfortunately; the trees protected from the herbivores became weaker and were actually more likely to die than those left open to the ravages of wandering leaf nibblers. It seems that the trees' defense system, composed of tiny ants living in their hollow thorns and feeding on nectar produced by the plant, had started to abandon the plot when they were no longer needed and the plants reduced their nectar production. They were replaced by another species of ant that allowed other insects, including a nasty boring beetle, to attack the tree. This simple example illustrates the complex and unpredictable web of interactions among living organisms in a relatively well-understood ecosystem, and the dangers of trying to modify one component without taking into account the effect on others.

We know even less about the complexities of human biology, although we are learning at an incredibly fast pace. Still, the biggest gap in our knowledge of ourselves is how the genome is translated into a living system. How do genes get turned on and off at the right levels to produce the biochemical mix in our cells, and how do these cells know how to work together to produce tissues, and how do tissues regulate themselves to create a viable organism? Then there is the question of consciousness, which is still not understood. Overall, the system is so incredibly complex that the mind boggles at how it all manages to work together. The answer is that billions of years of evolution have optimized the functions of all of these components until they work in harmony most of the time, except in rare cases where severe diseases result from seemingly trivial changes.

All of this complexity has led many people to question the ultimate utility of human genetic modification. While in theory many things may be possible using these techniques, for most people the risks will greatly outweigh the benefits. The bioethicist Leon Kass has written about what he calls "the wisdom of repugnance," where there is a subconscious sense of what is right and wrong in cases of biological modification. Although he first applied it to a discussion of human cloning, it could equally apply to other genetic modifications. Kass's argument has been criticized by many in the scientific community as a Luddite response, since all sorts of things that once seemed repugnant to many people—divorce, abortion, homosexual unions—are now commonplace. But as a brake on the unfettered application of all new scientific discoveries, Kass's guideline is perhaps a useful initial litmus test. Ultimately, though, the question of whether to apply these techniques should be part of a society-wide debate, not simply left up to physicians and their patients, egged on by pharmaceutical companies. Society, in turn, needs to become better educated about the risks and benefits inherent in such technologies. In a world where even physicians can make statements like "Charlie Whitaker should simply accept his fate," fewer than half of American adults can define DNA, and fewer than half of U.S. adults accept evolution, scientists clearly have a lot of work to do in communicating scientific concepts to a broader audience.

Overall, the debate should be about not what we are *capable* of doing but what we *should* be doing. As we move ever more quickly into a future that is a far cry from our origins as a species, we should take care to preserve what is important about being human. The most obvious choice in the twenty-first century may be to "pop a pill"—including one that changes your DNA—but we should always try to bear in mind what the long-term consequences might be. Certainly, we are clever enough to develop any number of new tools, including medical interventions, but are we wise enough to use them properly? There is a clear danger inherent in modifying the genes of the next generation, or our own, when we still don't understand enough about interactions and risk. Short-term thinking may lead us down a path that seems to offer infinite hope at the time but ends up with a nasty sting in the tail hundreds of years later.

Although the genetic possibilities in this chapter may still seem quite futuristic, and it will take us many generations to assess their ultimate impact, another such scenario is now playing out beyond biology. Decisions made nearly two centuries ago about fuel sources, egged on by our increasing appetite for energy, have led to an unprecedented impact on the natural world. Humans, armed with a desire for novelty and mobility, have used fossil fuels to take a ride. And now these same fuels may be returning the favor, in the form of a human-induced change in climate unlike any the world has seen in recent memory. Like the ice age returning at the onset of the Younger Dryas, and all that it meant for shifts in human behavior, climate change may already be impelling us toward a destination that will challenge what we think we know about human society and the world we live in. And as a model for the unquestioning acceptance of new technology, the story should serve as a warning for powerful technologies such as genetic engineering. Next stop: the South Pacific, and the leading edge in the battle over global warming.

## Chapter Six
# Heated Argument

*It is not the strongest of the species that survives, nor the most intelligent, but rather the one the most adaptable to change.*

—ATTRIBUTED TO CHARLES DARWIN

### TUVALU

The silence on the plane was almost deafening, psychologically blocking out the drone of the twin propellers. I was crusty with dried sweat, uncomfortable from having spent half a day standing on boiling-hot tarmac, waiting for the pilots to get the engines started on the fifty-year-old Convair aircraft. Without an engineer present, they had had to call New Zealand via satellite phone for step-by-step instructions on how to fix the problem. Finally, after tinkering for hours (during which we had wandered back and forth from the airport to the bar in the country's only hotel), the pilots tried one last time to turn over the right engine before the battery died. It worked, and everyone exhaled in unison. We piled back onto the plane with less than an hour to go until sunset, pleased to be getting out before—because there were no runway lights—darkness grounded us for the night. Each of us worried for a few minutes about whether we would make it the two and a half hours back to Fiji across a huge stretch of open ocean—no emergency landing strips if anything went wrong—but then settled in for the trip.

The reason for the silence, though, was not worry about the safety of the aircraft. Rather, it was shock over the dead woman wrapped in blankets and strapped to one of the front seats. About half an hour after our takeoff from Funafuti Atoll, the main island of Tuvalu, she had collapsed in a heap on the floor. A team of Australian doctors who had been in Tuvalu for the past two weeks immediately leapt into action, attempting to resuscitate her for nearly half an hour. Unfortunately, despite their best efforts, she had been killed by a blood clot in the brain before her slumped body even left the seat. Halfway to Fiji, with no way to return to Funafuti because of the darkness, the crew covered her up and strapped her into her seat, and we continued on our way. Everyone sitting in the forward half of the plane was witness to the whole ordeal, and word quickly circulated among all of the passengers: a woman had died on board.

These events were even more sobering for me because of the ten days I had just spent in Tuvalu. A tiny country of nine coral islands about 650 miles north of Fiji, 8 degrees south of the equator, Tuvalu gained its independence from Britain in 1978 (though it's still part of the Commonwealth). Its population of twelve thousand ekes a living from some of the most unforgiving land on earth, supplemented by royalties from a fortuitous Internet domain name: .tv. Tuvalu was settled by Polynesian seafarers around 3,000 years ago, and its people historically practiced pit taro cultivation, in

**FIGURE 27: FUNAFUTI ATOLL, TUVALU.**

which holes are dug into the coral stone of the atoll and filled in with organic material (pig feces, trash, palm fronds, and other detritus); the resulting "soil" is used to grow the starchy tuber that until recently provided the majority of the Polynesians' calories. Of course, the reefs are rich in marine life and have always yielded a bountiful fish catch. I had the chance to experience this myself during my visit, when I spent an afternoon spearfishing with a Tuvaluan near his family's *motu,* a small, sandy island across the lagoon from the capital, Funafuti. Practically everything else in the Tuvaluans' lives, though, is meager and hard-won, particularly fresh water. Tuvalu is truly a country that lives on the edge.

The reason I had come to Tuvalu was to see how close to the edge it really is. It is one of the lowest countries on earth—no point is higher than fifteen feet, and most of the islands are only a few feet above sea level. As a result, much of it is forecast to disappear beneath the waters of the Pacific by the end of the century, inundated by rising sea levels as the earth's climate warms. Long before this, storm surges and high tides will have gradually, inexorably killed the ancient taro pits with salt water and rendered the freshwater supplies undrinkable. Eventually agriculture will be impossible, and everything in the Tuvalu diet apart from fish will have to be imported. At that point, the Tuvaluans' way of life, so inextricably linked to the sea, will have been destroyed by it.

Tuvalu is not alone in this challenge. Every reader of this book is likely aware that the earth's climate is changing. Whether you've seen Al Gore's influential 2006 film *An Inconvenient Truth* or are a long-standing student of carbon credits and alternative energy supplies, climate change is one of the most pressing and best-publicized dilemmas facing the world in the twenty-first century. How radical these changes are remains to be seen, and estimates of the extent of global warming range up to ten degrees Fahrenheit by the year 2100. As the climate shifts, much of the polar ice that has been in place for millions of years will melt, raising sea levels by anywhere from three to six feet—enough to doom Tuvalu. Global sea levels rose by around eight inches in the twentieth century, and much of this was in the latter half, as global warming accelerated.

Tuvalu, unlike other low-lying regions threatened by inundation, has taken the matter to heart in a radical way. In 2002, sensing the impending loss of their homeland in less than a century, Tuvalu's leaders drew up plans to file a lawsuit against the United States and Australia in the International Court of Justice for their failure to ratify the Kyoto Protocol on climate change. They have also started plans for a mass evacuation of every citizen to New Zealand and its South Pacific dependency Niue. That's right—an evacuation of every person currently living on the islands. The details are still being worked out, but such a plan is unprecedented in world history.

"Climate refugee" is a term that will become much more widely known in the twenty-first century, as our planet goes through a period of radical climatic change unlike any since the end of the last ice age. And as with that period, and its resulting era of innovation, this new climatic shift promises to result in drastic upheavals to our way of life that are perhaps unimaginable today—including the evacuation of entire countries.

## KYOTO'S BLEEDING EDGE

The flight into Tuvalu had been much less eventful, though a storm about halfway there had caused me to wonder about the advisability of making the trip. Tuvalu gets only a few dozen tourists each year, and most of the people on the flight with me were either aid workers or Tuvaluans returning from shopping or business trips to Suva, Fiji's capital city. As we descended ever closer to the waters of the Pacific, I peered anxiously out the window, hoping to catch a glimpse of land. Apparently my furrowed brow was noticeable, because a Tuvaluan sitting across the aisle told me not to worry; we weren't going to land on water, he said, before laughing with his girlfriend. If this was the direction Tuvalu was headed in, it was reassuring to hear that it hadn't happened yet.

Stepping onto the tarmac, I was hit with a blast of steamy tropical air that reminded me just how close to the equator I was. Suva had been cold and rainy when we left, but here on Funafuti—the largest and most densely populated of Tuvalu's eight inhabited atolls—it was sunny and very warm. I immediately broke into a sweat, straining under the weight of my backpack in the intense sun. Global warming indeed, I thought, and headed over to the passport control in a wooden building at the edge of the runway.

Around 150 people were gathered outside, waiting for relatives or meeting the few foreigners who were arriving for development projects. More were on the runway itself, off to the side, having been cleared off just before our arrival by a car with a siren. Steve, an Australian legal-aid worker based on Funafuti, told me that the runway was the *de facto* center of social life on the island; sports, socializing, exercise, and simply watching the arrival of the twice-weekly flights from Suva—it all happened here or nearby.

The runway on Funafuti had been built by the Americans during the Second World War. Tuvalu served as a base for bombing missions to Japanese-occupied Kiribati ahead of the Battle of Tarawa, and the runway was more than long enough to land the hulking B-17 Flying Fortresses—and long enough, certainly, to land a jet, if Tuvalu were ever to make it onto the tourist map.

Tuvalu certainly deserves to be on that map, I thought as I walked the hundred yards to the Vaiaku Lagi Hotel. Swaying palm trees, sun, a beautiful (but unfortunately increasingly polluted) turquoise lagoon, and friendly people would certainly be draws— a place where life seems decades removed from the twenty-first century, even if the hotel recently got satellite television and Tuvaluans make much of their foreign exchange from the Internet domain. I was looking forward to my week here.

After checking in I went down to the bar to watch the sunset and discovered that it was the main center of expat life in Tuvalu. Over a glass of Fiji lager I chatted with a German named Wulf and his Fijian girlfriend about what had brought them here. He was working on a geospatial mapping project, creating a high-resolution map of the islands so the government could define exactly where its territory began and ended. But they had been in Tuvalu for too long, they told me, and were ready to head back to the "big smoke" of

Suva, where she could party in the style she preferred. I also met an Australian nurse who had come as part of the team that I would later fly out with. Its doctors were performing much-needed *pro bono* cataract surgery for two weeks on a population that spent its life in the intense tropical sun. She told me that both the hospital and a nearby desalination plant had been donated by the Japanese government in exchange for whaling rights in Tuvalu's waters. Though I'm no fan of whaling, I had to hand it to the Tuvaluans for their ingenuity in attracting foreign aid. However, she explained, although the hospital was brand-new, there were not enough trained staff members or supplies, so the Tuvaluans still had a shockingly poor standard of health care. That was why she and her team were there.

Gilliane and Chris showed up after I had been at the bar for an hour or so. They were filmmakers who had come to Tuvalu a couple of years before to make a documentary on its environmental situation (it's entitled *Trouble in Paradise* and is a very good overview of the situation facing Tuvalu). We all ordered dinner from the restaurant, and they told me about the project they were starting on the island: an environmentally sustainable power plant using pig waste. The problem was, they explained, that the Tuvaluans had been seduced by the same things as those of us in the West— cars and generators among them—and they were now net contributors to the global carbon problem. This made it difficult for Tuvaluans to argue that other countries were thrusting environmental devastation upon them, but as a small island nation they had the opportunity to choose a more environmentally friendly route into the twenty-first century. I was fascinated by the pig-waste project, and Gilliane and Chris promised to take me to the nearby island where they were building their first facility.

After chatting for a couple of hours, with a stunningly beautiful sunset as a backdrop, and eating dinner, I headed back to my room, tired but excited about my next week on the island. The remoteness of the location was tempered somewhat by the international mix of people I had just met—German, French, American, Australian, and Fijian, all brought to Tuvalu to help it or tell its story.

This diversity hammered home to me just how interconnected the world had become, and made me feel that Tuvalu did have some hope of survival after all.

Over the subsequent few days I met several other aid workers (there were no tourists at all when I was there), government officials, and many local Tuvaluans. Semese Alefaio was managing the Tuvalu Association of NGOs but had previously worked for the Funafuti Conservation Area—created by Tuvalu in 1996 "for the benefit of the community and future generations." He wanted to create a nongovernmental organization devoted to revitalizing the ways of his ancestors, like rainwater collection with palm fronds, or the harvesting of coconut gel for a baby's first solid food at six to eight months. Jeff, a Tuvaluan who had been living in Singapore, had come home to run in the parliamentary elections. He had lost and was now planning to return to Singapore, but had been prepared to stay if he had won. I was beginning to get the impression that there were a fair number of people who hadn't given up on Tuvalu.

On my last morning there, I met with Enate Evi, director of the Tuvalu Environment Department. He told me about the massive shifts in the Tuvaluan economy over the previous decade, a change from a subsistence lifestyle to a cash economy and imported food. Food imports brought with them a huge increase in waste from the packaging, which was currently rotting at one end of the island, and he discussed his plans to recycle aluminum cans and other items. We talked about the tidal surges and how they were making the taro pits too salty for cultivation, as well as the droughts that were becoming more frequent, especially on the northern islands. These factors were leading to more immigration from the outlying islands to Funafuti. Faced with few job prospects and a growing population, some young people on Funafuti—those who knew a trade and were willing to completely remake their lives—were emigrating, but there was no official large-scale effort to resettle the Tuvaluan population in another country. Evi told me of his desire to return to renewable energy sources (solar had once been much more widespread, but now a generator was easier and cheaper) and spoke highly of the work Chris and Gilliane were doing. He said that the much-discussed lawsuit against the United States and Australia had been dropped as too expensive and ultimately untenable, and he now saw the only way forward as being one of partnership with the larger countries, especially those in the European Union and, perhaps, the United States.

Overall, I came away from Tuvalu with a profound sense of a people in flux, trying to cope with their increasingly dire situation by thinking transgenerationally, considering more long-term goals, renewable resources, and transnational collaboration. This is a vision of the future that many in the developed world share and is, I believe, the only way forward in dealing with the enormous pressures of global climate change. As I write this I am looking at a glass Coca-Cola bottle that Semese and I found on the *motu* during our day of exploration. It was left behind by American soldiers stationed in Tuvalu during the Second World War, more than sixty years before. It still looks brand-new, and seems a fitting example of the sort of transgenerational power I've written about in this book. I doubt the young soldier who drank the Coke and tossed the bottle into the bushes had any idea that it would find its way—possibly long after his death—to a suburban bookshelf in Washington, D.C., a souvenir from a small island nation that is struggling to cope with the challenges of the twenty-first century.

## CAP IN HAND

The situation in Tuvalu demonstrates some of the innovative ways that people with very little material wealth or technology are dealing with climate change. As I've traveled around the world for my work, I've had a chance to see other options being explored. The key point, ultimately, is that people are reacting to the challenges that we face as a species. Our lifestyles, built on expansion and on using all of the resources available to us, are in the process of shifting to take into account the long-term conservation of those resources.

While I was writing this book, the price of a barrel of oil hit an all-time high of nearly $150, the most expensive it has ever been. The cost of buying gasoline became prohibitive for some of the rural poor in the United States, and many were forced to stop driving their own vehicles, move closer to their jobs, take public transportation, or carpool. Some of the lofty price of petroleum can perhaps be explained by speculation on the commodities markets or the decline in value of the dollar, but much of it is due to the increasing demand from China, India, and other developing countries as they have become richer. Although oil prices are extremely volatile, dropping to around $30 a barrel only a few months after hitting that all-time high, the fact remains that petroleum is a scarce commodity that seems destined to become more expensive in the future.

China is now the world's largest market for automobile sales, having surpassed the United States in December 2009. Moreover, Chinese sales are accelerating, while those in the States and Europe are declining, a trend that pre-dates the recession of 2008–09. Some economists expect India to be the world's third largest car market by 2030. Increasing wealth brings with it increasing desire for mobility, and the car remains the transport of choice. Even as we in the "rich world" may wring our hands about our carbon footprints and gas mileage, consumers in China still have the pedal to the metal on the automotive front. This is in part because of gasoline subsidies, which kept prices in China at about half of what they were in the United States through mid-2008. These subsidies have now been reduced significantly, but the Chinese continue to buy more cars every month.

The effects of this transportation shift are already visible—quite clearly, as anyone who has visited Beijing or Shanghai recently can tell you. City thoroughfares, once a sea of bicycles, are now traffic gridlocks, with millions of cars spewing exhaust into a horribly polluted sky. Even on sunny days it is sometimes difficult to make out the scores of building cranes on the horizon because of the smog, and the air quality is consistently so poor that some elite athletes seriously considered skipping the 2008 Beijing Olympics.

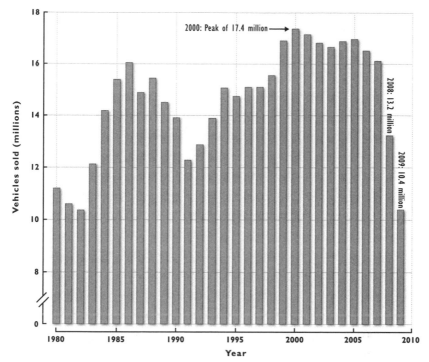

FIGURE 28: AUTO SALES IN THE UNITED STATES, 1980–2009.

In its race to embrace the fruits of the twenty-first century, China has also thrust itself into the heart of a debate about the downside of modernization.

The Kyoto Protocol doesn't address Chinese emissions standards. In fact, it fails to address emissions standards for any of the larger developing countries, including India and Brazil. They are exempt from the standards of the so-called Annex I countries, which are bound by the treaty to reduce their emissions of greenhouse gases— primarily carbon dioxide—to an average of 5 percent below 1990 levels. This is to be achieved through a "cap and trade" system, where Annex I countries are given a cap for their emissions levels and are able to trade "carbon credits" in order to help them achieve it. The trade in carbon credits recognizes that some countries—and even some industries within a country—will generate more green-house gases than others, and will be allowed to purchase the unused credits from the cleaner country or industry. In other words, the system tries to use both government-level quotas and the free market to achieve a net reduction in greenhouse gas emissions.

While almost all of the world's countries have ratified the Kyoto Protocol (it actually came into effect when Russia ratified it in 2004), the largest carbon emitter at the time of the protocol's 1997 drafting, the United States—which then accounted for around 40 percent of global carbon emissions—never signed. The reasons given by the administration of President George Bush were twofold. First, officials argued that the economic costs were too high as proposed and there needed to be more flexibility in implementation. This, perhaps, can be dismissed as protectionism and a shortsighted failure to recognize the long-term costs of global climate change. Second, and perhaps more important, the administration pointed out that it hoped to have emissions standards set for developing countries as well. Noting that China is building an average of one large coal-fired power plant every week, detractors of the protocol say that the stakes are too important to leave developing nations out of the reduction targets. In fact, in 2006 China overtook the United States as the world's largest carbon emitter. The Chinese counter that they not only should benefit from the same sort of petroleum-assisted development that took place in the world's richer countries in the twentieth century, but also that, as a nation of 1.3 billion people to the United States' 300 million, their emissions per capita are still much lower.

As I write this in early 2010, the battle still rages, but the election of Barack Obama has yielded a significant shift in U.S. climate-change policy. The administration, while still not ratifying Kyoto, is focusing on the proposal drafted at the U.N. summit in Copenhagen in December 2009 to draw up a successor to Kyoto, which is set to expire in 2012. With America's new willingness to "play ball," the Chinese seem to be more willing to contemplate capping their own emissions. The debate is clearly far from over, however, and even well-intentioned European countries like the Netherlands and Austria are falling well short of their agreed emissions levels. Whatever happens, it is clear that China, India, and Brazil will ultimately need to be put on a carbon diet if the treaty is to have a long-term effect on global warming.

## A HEATED ARGUMENT

The debate about how to reduce global greenhouse gas emissions is based on some very complex science. It was only in the past decade that the scientific establishment even fully accepted the importance of carbon dioxide levels to global temperatures, culminating in the release of a landmark report by the Intergovernmental Panel on Climate Change (IPCC) in 2007. This document, put together by hundreds of scientists and policy makers from around the world after widespread scientific consultation, came to the verdict that the global climate was unequivocally warming and that humans were responsible for much of the temperature increase in the twentieth century. Despite a scandal involving leaked emails between some of the scientists involved in drafting the report, and an incorrect estimate of when Himalayan glaciers could disappear, it remains an important overview of the state of climate-change research.

The underlying patterns that led the IPCC to issue these statements are fairly straightforward, and really come down to two observations. The first is the unprecedented increase in atmospheric carbon dioxide over the past century. Extrapolating from the study of gas bubbles frozen in deep polar ice, climatologists have calculated that the average carbon dioxide concentration in the atmosphere over the past 650,000 years has varied between 180 and 300 parts per million (ppm). The level at the start of the Industrial Revolution, in 1800, stood at around 280 ppm. In 2008 it was 387 ppm, far above the long-term average, and it is currently increasing at around 1.9 ppm per year. The implication is that the massive increase in the burning of fossil fuel since the dawn of the industrial age has added around 100 ppm of carbon dioxide (one of the products of fuel combustion) to the atmosphere. Carbon dioxide traps radiative heat from the sun, much like the panes in a greenhouse,

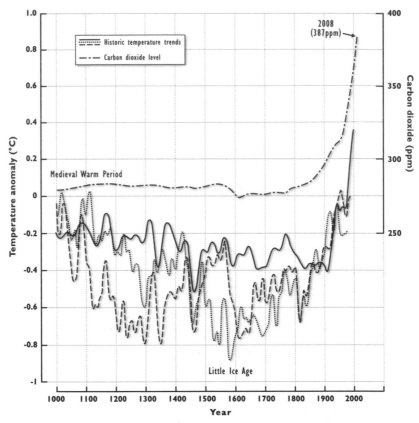

**FIGURE 29: PATTERN OF INCREASING CARBON DIOXIDE LEVELS AND TEMPERATURES OVER THE PAST 1,000 YEARS.**

with the net effect of warming the earth's atmosphere. Other greenhouse gases, such as methane and nitrous oxide, have also increased substantially during this time period, but carbon dioxide is the most important contributor to global warming.

The second pattern that underlies the IPCC's recommendations is the so-called hockey stick graph of Northern Hemisphere temperatures over the past millennium. Between A.D. 1000 and 1900 temperatures varied quite a bit from year to year, dipping slightly from the sixteenth through the nineteenth centuries (a period known as the Little Ice Age), then rising from the beginning of the twentieth century. Since 1950, though, temperatures have increased substantially over their long-term average, and the last two decades of the twentieth century were the hottest in more than

1,000 years. The hockey stick graph has attracted quite a bit of controversy, largely based on its modeling of temperatures prior to the past couple of centuries, and many consider it to be a gross oversimplification of complex historical climatic patterns. What seems clear from all of the models, though, is that global temperatures have increased in the past half century.

The effects are clearly visible, even without the climate statistics. The Arctic region seems to be the worst affected, with ice sheets melting in regions that used to be frozen throughout the year. Polar bears will almost certainly become extinct as their habitat disappears. Species that typically lived farther south are now more common in the far north. The 2008 Audubon Society Christmas bird count, for instance, documented species such as the purple finch and the wild turkey as much as four hundred miles farther north than their previously typical range. Glaciers are disappearing around the world, and coral reefs are increasingly showing signs of bleaching—a reaction to rising sea temperatures and acidity, as the mounting carbon dioxide is absorbed by the world's oceans.

Coupled with the increase in greenhouse gases during this time period, the implication is that humans are causing global temperatures to rise, primarily through the burning of fossil fuels. What is debatable is the degree to which humans alone are producing global warming. We have been in a gradual warming trend since the Little Ice Age, and some people have tried to argue that what we are seeing in the twentieth-century temperature record is simply part of a more long-term natural trend. The vast majority of scientists and global policy makers, however, now agree that we are the primary culprit. The recommendations of the IPCC underlie the creation of Kyoto and other planned treaties aimed at curbing carbon dioxide emissions, and whether they achieve their stated goals or not, they will continue to be a strong influence on global energy policy for decades to come.

Much has been made in the press about the predictions contained in the IPCC report, which, using projections of greenhouse gas emissions, attempt to extrapolate from the existing data and computer models to estimate future temperature and sea level rises. This is currently the cutting edge of the scientific debate about global warming, and much remains to be worked out in our understanding of the complex forces we have set in motion. This short chapter is clearly not the venue for a detailed discussion of these predictions. One small point in this section of the IPCC's report has not been as widely discussed, however, and it is perhaps the most chilling—at least when viewed from the perspective of a layperson, rather than a climate-change wonk. This statement from the IPCC "Summary for Policymakers" puts it clearly and succinctly:

> Both past and future anthropogenic carbon dioxide emissions will continue to contribute to warming and sea level rise for more than a millennium, due to the time scales required for removal of this gas from the atmosphere.

This means that, whatever we do to mitigate global warming by setting caps and trading carbon credits, the forces we have set in motion will radically change the world we live in over the next millennium. As with the power of genetically modifying our offspring and, before that, domesticating plants and animals during the Neolithic period, we have set in motion transgenerational forces whose ultimate effects we simply cannot predict. Although much of the debate around global warming centers on models and predictions, with science at the fore, the fact that we have started something that will affect us in some form or another is a certainty. Global warming will probably be the biggest social challenge of the twenty-first century—and perhaps beyond—as we adjust to these climatic changes. And if history is a guide, these changes will be profound indeed.

## THE IMPORTANCE OF SUMMER

Today, the three tallest mountains in Oceania—which includes the Malay Archipelago, Australia, New Guinea, and the Pacific islands—

are found on the islands of New Guinea and Borneo. The Carstensz Pyramid, at 16,023 feet, is the highest, one of the Seven Summits fetishized by mountaineers hoping to climb to the highest point on every continent. Puncak Trikora, formerly Mount Wilhelmina, sits nearby at 15,583 feet, while Mount Kinabalu, on Borneo, is 13,435 feet. Until 1815, however, Kinabalu was in fourth place. It rested in the metaphorical shadow of Mount Tambora, the highest point on the small island of Sumbawa, just east of the tropical paradise of Bali. Reaching an estimated height of more than 14,000 feet, Tambora's huge, symmetrical cone soared out of the Flores Sea, which separates Sumbawa from Sulawesi. Navigators had used it as a landmark for thousands of years, and it was believed by the local people to be the home of their great god.

According to local legend, the events that reduced Tambora to a fraction of its former height happened as a result of divine retribution:

> *The cause was said to be the wrath of God Almighty*
> *At the deed of the King of Tambora*
> *In murdering a worthy pilgrim, spilling his blood*
> *Rashly and thoughtlessly*
>
> Syair Kerajaan Birama,
> TRADITIONAL POEM, QUOTED IN BERNICE DE JONG BOERS,
> "MOUNT TAMBORA IN 1815,"
> *Indonesia* (VOLUME 60), OCTOBER 1995

On the evening of April 5, 1815, Mount Tambora, a dormant volcano, erupted. Actually, "erupted" is an understatement; it exploded, it roared, it destroyed, it belched forth demons. The sound was heard in Yogyakarta, 450 miles away, where the sultan rallied a detachment of soldiers to see if the capital was under attack. Over the next several days the eruption continued, reaching a climax on April 10, when the sound was heard in northern Sumatra, more than 1,500 miles away from Tambora. Sir Stamford Raffles, the colonial founder of Singapore, who gave his name to the famous hotel, described it in his memoir from eyewitness accounts:

In a short time, the whole mountain . . . appeared like a body of liq-
uid fire, extending itself in every direction. The fire and columns of
flame continued to rage with unabated fury, until the darkness
caused by the quantity of falling matter obscured it at about 8 p.m.
Stones, at this time, fell very thick . . . some of them as large as two
fists.

Ash was thrown more than twenty-five miles into the sky, producing three days of darkness in the nearby islands. Perhaps the
most powerful volcanic eruption since that of Mount Toba 74,000
years before, it spewed out more than twenty-five cubic miles of
ash and stone, four times as much as its more famous Indonesian
cousin, Krakatoa. The island of Bali, 150 miles to the west, was
covered in nearly a foot of ash, and winds spread more ash worldwide over the next few months. The eruption reduced Tambora's
lofty elevation to today's relatively paltry 9,350 feet—a significant
drop in stature. It is thought that more than seventy thousand In-
donesians perished as a result of the blast and its aftereffects, making it the deadliest volcanic eruption in history.

Despite the catastrophic effects of Tambora's eruption on Indonesia, the most long-lasting effect had little to do with its
Hollywood-style pyrotechnics. Rather, it was the hundreds of millions of tons of sulfurous gas ejected into the stratosphere during
the eruption. This, coupled with fine particles of ash blown high
into the atmosphere, produced a haze that was seen as far away as
London and New England. Although it resulted in some spectacular sunsets, its long-term effects were to be far more insidious.

As outlined in their fascinating book *Volcano Weather: The Story
of 1816, the Year Without a Summer,* Henry and Elizabeth Stommel
described how the reduction in the sun's intensity caused by Tambora's stratospheric sulfurous belch created a global cooling effect
the following year, 1816. The fact that the volcano was located
practically on the equator made it easier for wind patterns to carry
the effluent around the entire world, and the effect on the weather
was pronounced. Amateur observers—including several meteoro-

logically minded Ivy League scholars—who had been keeping careful records of daily temperature readings for many years noticed a pronounced drop in June of that year. Local farmers saw the effects directly: a cold front that moved through New England between June 6 and 11 left several inches of snow on the ground as far south as Massachusetts and the Catskills of southern New York State, and a hard frost occurred in Pennsylvania and Connecticut. Two more frosts arrived in July and August, effectively destroying the New England corn harvest that season, despite attempts to replant after each cold front dissipated.

Things were even worse in Europe. An exceptionally cold and wet summer in 1816 battered wheat and corn crops in the northern countries. The resulting famine contributed to a typhus epidemic in Ireland, and thousands died of hunger in Switzerland. France, having been defeated the year before at the Battle of Waterloo, erupted into riots over the high price of grain. Europe's problems would continue until the middle of the following year, when somewhat better temperatures allowed a harvest that was closer to normal. Overall, historians estimate that as many as 200,000 Europeans may have died as a result of the cold weather in 1816.

Perhaps the most fascinating results of the cold snap that summer, though, were social. Most New Englanders were subsistence farmers, and the loss of their crops hit them very hard. Lured by the prospect of easier lives far removed from unseasonal frosts and harsh winters, many packed up and moved out, headed for the Midwest. The Stommels compare the emigration to the exodus during the Dust Bowl era of the 1930s, though that, being spread over several years, was more extreme. It is clear, however, that the freakish summer weather forced many people over the edge and led to a huge population increase in places like Ohio and Indiana— American Manifest Destiny helped along by a volcanic eruption in Indonesia.

Other, lesser social changes may trace their origins to the Year Without a Summer. A *New Scientist* article published in 2005 argues convincingly that the velocipede, the ancestor of the modern bicycle, was developed in the aftermath of 1816's failed grain harvests as a means of transportation that didn't need to be fed (unlike horses). And in perhaps the most unusual effect of the poor weather, Mary Shelley took refuge from the elements in a Swiss castle being rented by Lord Byron, using her time to write the horror classic *Frankenstein.*

A cautionary story that has lent its name to a term of disparagement used by the anti–genetic modification movement ("Frankenfoods"), a means of transportation now used by more than a billion people around the world, and an America political ideology rooted in nineteenth-century westward expansion—not, perhaps, what you would expect from the aftermath of a summer's worth of bad weather. What these disparate outcomes illustrate, though, is the unexpected long-term effects of climate changes, even relatively small ones that last less than a year. When we compare this to the much more extreme changes that are coming our way in the next century—whoever's model turns out to be correct—it is truly frightening. But as with Pandora's captured spirit, there is a spark of hope amid the gloom.

## DARWIN'S IMPERATIVE, OR THE GERMAN SOLUTION

During 2008 and 2009, while I was thinking about and writing this book, economists and policy makers around the world were reeling from two financial punches. The first, a credit crisis spurred by falling property values in the developed world, threatened to kill investment by making borrowing money more difficult. Without investment, global capitalism is left stranded on the side of the road—it is the energy source that drives the economic engine of modern society. Artificial stimuli such as lowered interest rates and tax rebates may provide a short-term burst of power, but this is simply borrowing by another name—borrowing from future tax revenues to pay for today's crisis. Some believe that it actually amounts to the greatest transfer of wealth in history, as the costs of patching today's leaky economic boat will be borne by the taxpayers of the future. Ultimately, the economic engine will need to slow down while investors count their losses and begin to determine where to invest more wisely. As with all of the classic bubbles of the past four centuries, from tulips to South Sea Company shares to the dot-com boom of the 1990s, what Alan Greenspan has called "irrational exuberance" will need to give way to rationally based investment decisions. Though painful, such corrections seem to be a periodic feature of a capitalist system, and others will surely happen in the future.

The second financial crisis is more serious, and somewhat unexpected. Prices for commodities—those tangible goods used to create the things that we consume—skyrocketed in the late 2000s. Oil, coal, and natural gas roughly doubled in price between mid-2007 and mid-2008, as did rice, corn, wheat, and milk. As a result, governments around the world were hit with unrest. Most of the world's food calories come from these three grains, and the vast majority of its power comes from fossil fuels, so the effects of higher prices are particularly difficult, especially for the poor. In February 2008, thousands of Mexicans protested a 400 percent increase in the price of tortillas, their dietary staple, and Haiti's government was toppled by rioters incensed at the high price of food and fuel. In June 2008, hundreds of Spanish truck drivers blocked highways, demanding lower fuel prices. Americans with SUVs winced as the cost to fill up their gas tanks rose to the triple digits. These price increases, while bad enough on their own, also raised the rate of inflation across the entire economy. Since fuel powers everything we do, the subsequent effects of higher prices were felt everywhere, from more expensive air travel to higher home electricity bills. Although oil prices dropped precipitously as the global recession took hold during the latter half of 2008, the fragile recovery of late 2009 sent it climbing toward $100 a barrel again.

The reason this surge in commodity inflation is a surprise is that commodity prices had, until recently, been *falling* in inflation-adjusted terms. In a now-famous wager, Paul Ehrlich, the influential author of *The Population Bomb,* bet the economist Julian Lincoln

Simon that commodity prices (in particular, a selection of metals) would rise in the 1980s. In fact they fell, and Simon won the wager. Similarly, food prices fell throughout the latter part of the twentieth century as the agricultural advances of the Green Revolution took effect. All of this occurred despite an increase in population of 2.5 billion between 1970 and 2000. It seemed that humanity could have its cake and eat it too—keep on expanding and pay less for the basic staples of life.

Looking past the temporary reductions associated with a recessionary period, the era of falling commodity prices now appears to have come to an end, at the same time that global warming is entering the world stage. It seems that humanity is on a collision course with disaster, a planetary *Titanic* hell-bent on destroying itself. Doom-and-gloom prophets speculate on whether we will even survive as a species. It's all so depressing . . . or is it?

During the First World War, armies around the world manufactured the gunpowder used for ammunition using a slightly refined version of a centuries-old method. Combine charcoal (carbon), saltpeter (potassium or sodium nitrate, a naturally occurring source of nitrogen), and sulfur (this was not used by the early twentieth century) in a particular ratio and voilà: a mixture that burns rapidly and explosively when ignited. The problem was that while charcoal was easy to obtain, most of the world's saltpeter came from sources halfway around the globe, particularly in Chile. During the war, Germany was cut off from these supplies and had to find another source or face defeat. As it happened, in 1910 a German chemist named Fritz Haber had developed a way of manufacturing ammonia (which also contains nitrogen) from gases in the air. By scaling up this method, known as the Haber-Bosch process, German workers were able to produce a substitute for Chilean saltpeter and continue the war effort. Similarly, when subject to Allied fuel blockades in the Second World War, the Nazis developed a way of manufacturing a gasoline substitute from coal. In both cases, the old maxim about necessity being the mother of invention proved true.

The point of the anecdote is not to illustrate how clever the German war machine was in these two devastating conflicts but to show that circumstances can dictate when change is necessary. If the cost is high enough, it pays to innovate. Humans, uniquely among all of the earth's species, have developed cultural inventions that allow them to adapt to virtually any circumstance. No oil? Make it from another source. Want to walk on the moon? Build a rocket to take you there. Too many people to feed? Breed higher-yielding crop varieties and create government policies that reduce the population growth rate. The more intense the crisis, the greater the incentive to develop a solution.

Now, as we move into the twenty-first century, we are facing another period of crisis. The last episode of intense climate change we experienced as a species, during the Younger Dryas, led a few hunter-gatherers living in the right locations to start cultivating crops, which enabled all of the subsequent technological innovations of the Neolithic era. These included significant genetic changes to the crops themselves, the domestication of animals, the development of complex irrigation systems, and the rise of urbanism and multilevel government. All were developed in response to changes ultimately set in motion by the climatic shifts at the end of the last ice age. The question facing us now concerns what will be developed in this new era of climatic change—whether you believe that humans are responsible or we are simply contributing to a longer-term warming trend. In other words, what is the opportunity in this crisis?

It is clear that modern humans live an incredibly energy-intensive lifestyle. Americans use an average of more than 270 kilowatt-hours (kWh) each day, while people in India and Africa average just over 10—the energy content of a liter of oil. Not surprisingly, hunter-gatherers use even less, perhaps under half of this amount. As a species, though, we have been in an accelerating trend of energy consumption since the dawn of the Industrial Revolution. This has brought us many of the conveniences that we cherish in modern life—particularly mobility, as around 40 percent of our energy consumption goes to power automobiles, airplanes, and the transport of goods. Authors and politicians bemoan our "energy addiction," but it is unlikely to go away anytime soon. So are we like heroin addicts, heading toward rock bottom as we burn through the last of our hydrocarbon sources in the next century, oblivious to the carbon emissions they are spewing into the atmosphere?

Fortunately, thanks to the laws of economics, the answer seems likely to be no. As with other crisis points in the past, once the costs become great enough, we begin to look for alternatives. Like the Natufians forced to roam farther and farther from their permanent encampments in the Fertile Crescent 12,000 years ago, we are beginning to feel the costs associated with using a nonrenewable resource like hydrocarbons. This is the thinking behind Kyoto and other proposals for taxing carbon emissions: to assign an environmental cost to the burning of fossil fuels. It seems, too, with the age of what has come to be called "peak oil," that the marketplace may be stepping in to regulate what governments cannot.

"Peak oil" refers to the point at which oil is maximally available from any particular source—when it is at its most plentiful. First proposed by the geologist M. King Hubbert in 1956, it accurately predicted that United States oil production would peak in the late 1960s. Supplies in other countries have followed suit, and even Saudi Arabia may have passed its peak. Meanwhile, world demand for oil is increasing—particularly in the developing world, where India, China, and others are speeding toward higher levels of oil consumption. Inevitably, according to the law of supply and demand, the price of oil must increase.

New sources may come online soon, and there are so-called unconventional reserves—tar sands, oil shale, and the like—but these are far more expensive to exploit than the giant underground petroleum bubbles that have fueled the industrial explosion of the twentieth century. The new hydrocarbon economics leave us with two possibilities: either we change our lifestyles radically or we find some other way to power our energy-guzzling society.

Several technologies are currently vying for a role in the new world of "alternative energy," as nonhydrocarbon energy sources are called. The chemical bonds between the atoms in petroleum compounds are simply nature's way of storing energy, like batteries. Plants living millions of years ago captured energy from the sun, and when they died some of it was trapped in ancient swamps and bogs. Over time, under the influence of intense heat and pressure deep in the earth's crust, these stored organic compounds underwent complex transformations into the substances we know as petroleum and coal. They are the most energy-rich naturally occurring substances on earth; as Thomas Homer-Dixon points out in his compelling book *The Upside of Down*, "when we fill our car with gas we are pouring into the tank the energy equivalent of about two years of human manual labor." It's no wonder that we're so addicted to them—nothing else even comes close in the energy sweepstakes.

The irony is that the original source of the energy in hydrocarbons is still ubiquitous and free: solar energy. While we may still have a way to go in developing high-efficiency solar collectors, the way forward seems clear: improve our ability to harness the ultimate source of energy on earth, the sun. Whether through photovoltaic cells, concentrating solar power (in which mirrors are used to focus the sun onto pipes that generate steam for electrical generators), wind farms, or wave-powered generators, all of the energy we hope to collect ultimately traces back to the sun and its powerful stream of free photons barreling down on us every day.

What even the most ardent supporters of solar energy admit, of course, is that it alone cannot power our current consumption patterns, at least not yet. In fact, even a combination of solar and other renewable energy sources wouldn't allow us to keep up the highly mobile, energy-guzzling lifestyles those of us in the developed world enjoy today. There are two solutions to this problem: use less energy or find another source. Energy efficiency is improving all the time—with the exception of average gas mileage for America's automobiles, which in 2007 was actually less than a Ford Model T produced a century before—but it's unlikely that this will lead to

178

enough of a reduction in consumption to allow us to live entirely off solar and other renewables. You also need quite a bit of energy to manufacture solar panels and wind turbines, which reduces their green credentials somewhat.

One possible answer is to move one step up the food chain from solar—to nuclear, the ultimate source of the sun's power. Though power produced through the fusion of hydrogen atoms, which is what takes place in the sun, is unlikely to become a viable model of power generation here on earth anytime soon, the splitting of heavy atoms into lighter ones, or fission, has already been applied successfully for over half a century. France produces around 80 percent of its electricity through nuclear fission, South Korea and Japan over 30 percent, and the United States nearly 20 percent. Most countries, however, concerned about the dangers of another Chernobyl-like disaster, as well as the political difficulties of waste disposal (who wants to live near a nuclear waste storage facility?), have not been as pro-nuclear, and overall only around 15 percent of the world's electricity comes from nuclear power.

This looks set to change over the next century, as nuclear waste disposal methods become increasingly sophisticated and power plants become safer and more efficient. One of the most promising new technologies is currently being tried in China and South Africa, a variant known as the pebble bed reactor. Most of the immense bulk of a conventional nuclear reactor is actually devoted to the cooling system, which circulates water around the fuel rods and through a complex systems of pipes and towers in order to dissipate the heat absorbed from the fission reaction. The pebble bed reactor obviates the need for this complexity (and substantially reduces the size of the plant) by using an inert gas such as helium as the coolant, passing it around pebbles containing a mix of the nuclear fuel (e.g., uranium) and graphite. These pebbles generate the energy, and the heat is dissipated into the gas. The design is inherently safer than a conventional water-cooled design because the higher the temperature gets, the less power is generated, until eventually the fuel stops reacting altogether.

Other advances, such as hydrogen cells (generating and burning hydrogen as a fuel) and ocean thermal energy conversion (making use of the difference in temperature between deep and shallow water), seem to be less promising and have substantial technical hurdles to overcome. In the case of the former, the problem is how to efficiently generate the large quantities of hydrogen that will be needed—currently the best source is hydrocarbons obtained from oil. In the case of the latter, the amount of energy obtained is so small that it would merit application only in a very limited number of cases—on atoll islands, for instance. Similarly, geothermal power is useful in places like Iceland—a small country that sits on a gap in the earth's crust, allowing easy access to the hot magma that can be harnessed to power steam-driven electrical generators—but not in Kansas.

Advances such as the pebble bed reactor, improved battery technology allowing all-electric cars with reasonable power and range, more efficient solar panels, and wind turbines that rotate in lighter breezes promise to replace much of our current reliance on fossil fuels. As the price of oil and natural gas rises over the next several decades, it will eventually become cheaper to use these alternative forms of energy, speeding their adoption. It is possible to foresee a combination of all of them being used in varying degrees—solar panels supplying a substantial amount of power at the household level in sunny areas, backed-up wind power where it is easy to install turbines (in the midwestern United States, for instance), with nuclear providing the balance and generating enough electricity to recharge battery-powered cars and light machinery. Only aircraft and heavy machinery will likely need to use petroleum, because of their huge power needs, and even these may one day run on electricity.

While this scenario may sound like the naïve musings of an eco-nut to some, and surely leaves out the complexity of the transition from petroleum that will certainly take place gradually throughout this century, it does seem like the only sustainable solution long-term. Eventually the oil supply will run out, or will become so expensive to extract from the ground that alternative forms of energy will be the only logical choice. We are just beginning to sense the long-term costs of inaction on climate change and peak oil, and the ease of modern communication is increasing the rate of information transfer between far-flung places. In a tentative but perceptible way, we are coming to see ourselves as part of a global community, not simply as Americans or Indians. If the world is flat, as Thomas Friedman famously pointed out, its residents are also becoming more aware of their rather perilous shared future.

It is possible that the sticking point in the Kyoto negotiations—carbon-emitting countries in the developing world—could actually end up showing us the way forward with sustainable energy. Recent announcements by the Chinese government suggest a move in this direction. And in a world of expensive hydrocarbon fuels, perhaps it will be easier to implement alternative solutions in places with less of a deep, systemic commitment to a petroleum-based lifestyle. Like cell phones, which leapfrogged landline usage in the developing world throughout late 1990s and early 2000s because cellular networks were cheaper and easier to install and there was little competition from landlines, easy-to-implement alternative energy solutions may first find their greatest audience in the less developed parts of the globe. In Mongolia, I've encountered nomads who obtain all of their electricity from solar panels; there simply is no power grid to plug into on the steppe. Tuvalu is experimenting with a method of generating methane for cooking stoves from pig manure. Yes, it's a hydrocarbon, but at least it's more sustainable than shipping in natural gas from thousands of miles away. Necessity, to reiterate, is often the mother of invention.

The coming fuel and climate crisis may, in fact, yield an unparalleled era of innovation as a dividend, one unlike any other we've experienced. During this new era, we will start to innovate with the realization that we're all connected, that what we do "here" affects someone else "over there," as well as future generations. It is notable that President Obama chose as his energy secretary Steven Chu, a Nobel Prize–winning physicist known for his commitment to alternative energy research. If this and other appointments are indicative of a new resolution to channel research dollars into sustainable energy, it should have a huge impact on the development of these technologies.

## BACK TO THE SEA

Australia is the driest contiguous landform on earth. It is also the most urbanized, with around 90 percent of its population living in cities. Despite its vast size, which gives it a relatively modest population density of 6.4 people per square mile (the United States, in contrast, has 76, while Bangladesh has around 2,200), most people are actually packed into a narrow strip of land along the coasts, with one-quarter of them living in Sydney alone. The reason is the harsh climate of the interior, which makes life all but impossible without complex systems to bring in water from the coasts.

People have long been coastal dwellers, for many reasons. A steady supply of protein can be obtained from coastal resources— fish and shellfish—that can't always be matched inland. The earliest evidence of human exploitation of coastal resources comes from Eritrea, where around 125,000 years ago people were already eating enough shellfish to produce a pile of empty shells large enough to survive to the present day. It's not only reliable protein sources that drive people to the coast, though—as in Australia, water has long played an important role. Moisture from the ocean falls as rain on the windward side of coastal mountains, filling rivers and allowing plants—and humans—to thrive.

With the current trends in climate change, more and more people will be forced to leave behind unproductive land in continental interiors. Drought, famine, and disease, as well as the opportunities on offer in the world's cities, will serve as powerful incentives to migration. The world is becoming more urbanized every year, and more than half of us now live in cities, most of which are located in coastal regions. This massive migration to the sea promises to overwhelm already highly stretched water supplies, leading to a full-scale water crisis. California estimates that dwindling supplies, population growth, and climate change will lead to a massive water shortfall by 2020, one for which the government currently has no solution.

The earth has a finite supply of fresh water, and much of it is wasted. While California faces the specter of widespread drought and draconian water rationing within the next decade, developers in parts of the western United States continue to build golf courses in the desert. Suburban home owners water lawns that are completely alien to the arid environments where they are being grown. Aquifers are drained to support farms whose long-term sustainability is in doubt. The Colorado River, which used to empty into the Gulf of California, is now depleted of water long before it reaches Mexico, except in unusually wet years. This is not just a North American problem, though—China and India are draining their own aquifers to irrigate farmland for their growing populations. The Aral Sea, in central Asia, has lost 80 percent of its volume as water from the two rivers that feed it has been diverted for irrigating cotton fields. The Sahel region in Africa, an arid grassland just to the south of the Sahara Desert, is experiencing significantly more drought years than in the past, and climate projections suggest that the trend will continue.

The response to such worrying trends, at least for the western United States, might be to use less water. Clearly, conservation is important, but conservation alone can't provide enough water for everyone. Purifying wastewater, which is currently thrown away in most countries, is one possibility, but it is expensive and difficult to implement. Since much of the world's existing fresh water is tied up in the ice sheets at the poles, theoretically they could be tapped for human consumption. However, the transportation issues involved in this would be staggeringly expensive. Ultimately, new supplies must be found or people must be relocated to areas with more reliable water supplies. While the latter is virtually impossible to envision on political grounds, the former may be possible in some situations through the application of technology, making use of the huge water reservoir in the world's oceans. The only problem, of course, is how to get rid of the salt.

Desalination has a relatively recent history. Though it was suggested by Thomas Jefferson over two centuries ago, and was used in the nineteenth century to generate fresh water for steamship boilers when they were out at sea, it was not investigated seriously as a way to provide drinking water until the middle of the twentieth century. It is still incredibly expensive and uses enormous amounts of energy, which means that it has been widely applied only in relatively wealthy places like the Persian Gulf states or on large ships. Worldwide it accounts for only around 12 billion gallons of fresh-water production every day—less than half of the residential water consumption in the United States, and far less than the total amount used once you include industry and agriculture. Given that global daily fresh-water usage, including all industrial and agricultural consumption, is around 800 trillion gallons a day, the 12 billion currently produced by desalination plants is literally a drop in the bucket.

It is possible, though, that for domestic consumption in coastal cities, desalination could eventually contribute a substantial proportion of the total. Particularly as newer technologies are developed, including nuclear desalination (making use of the heat from nuclear power plants) and reverse-osmosis approaches (where salt is removed from seawater by using a permeable membrane), desalination could become more important. Add this to the increasingly urbanized world population, throw in water-saving technologies like low-volume toilets, low-flow showers, and wastewater recycling, and it is possible to envision a time when much of the world's population is using the limited fresh-water supplies in a more sustainable way.

The big culprits, of course, are agriculture and industry. It takes at least 1,000 gallons of water to produce a pound of beef and more than 500 gallons to produce a pound of rice. Better technologies, particularly genetic engineering to develop varieties of plants that require less water, could reduce these numbers significantly. Unfortunately, the legacy of the Green Revolution has meant that most seed companies have focused on creating herbicide-resistant strains to complement the expensive herbicides they sell—a wonderful business model, but not very wise in terms of natural resource utilization. Ultimately, even with these technologies, with a rising world population we are still likely to need far more water than we have easy access to. The only long-term solution, it seems, is to use less—something that runs against the grain of modern life.

This still leaves many people, such as the inhabitants of Africa's Sahel region, stuck in a precarious geographic trap. They will soon no longer have enough water to sustain even the frugal lives they lead now, as the Sahara expands unrelentingly to the south. Global warming, groundwater depletion, and a long-term drying trend in the region will all combine to cause millions to flee their villages. But where will they go? How will the rest of the country, or the other countries in the region, cope with the influx? According to the Red Cross, more people are now forced to leave their homes because of environmental disasters than war, and more than 25 million people worldwide can now be classified as "environmental refugees."

Not all of them are in the developing world, either. California may be eyeing a future of environment-induced migration, but Mississippi and Louisiana's Gulf Coast residents may have already experienced it. When Hurricane Katrina made landfall there in 2005, more than 1 million people were forced to flee their homes. Many have not returned—it's estimated that at least 100,000 of them are now living elsewhere. According to many climate scientists, as the earth's temperature rises over the next century and the gulf waters become warmer, strong storms such as Katrina, which reached Category 5 on the Saffir-Simpson scale just before making landfall, will become more frequent. Katrina could be a sign of things to come.

As with the people of Tuvalu, the earth's changing climate will soon have a profound impact on all of us. We can choose to respond by turning a blind eye, hopeful that the effects of global warming aren't as serious as they seem to be, or we can take the coming changes as a powerful incentive to remake both our technological and our social orders. The twenty-first century will test much of what we take for granted about our current lifestyles, and will force us to finally come to terms with the powerful transgenerational trends we unleashed during the Neolithic period. Climate change provides us with an opportunity in crisis, a powerful reason to change deeply entrenched behaviors. As a species that has long been accustomed to growth, expansion, and consumption, we will have to use our ingenuity in new ways to create a lifestyle with long-term sustainability. It's certainly not going to be an easy transition, but just as with the other crisis points in our species' history, we do possess the intellectual abilities to adapt. First, however, must come a sea change in our worldview.

# Chapter Seven
# Toward a New *Mythos*

*All gods are homemade, and it is we who pull their strings, and so,*
*give them the power to pull ours.*

—ALDOUS HUXLEY

## LAKE EYASI, TANZANIA

I spent the morning tracking animals with my friend Julius. As we
walked between the trees and across the open grasslands of the savanna, he showed me the tracks of elephants, impalas, porcupines,
and half a dozen other species, left like fleeting messages in the sandy
soil. Julius's ability to discern even the faintest evidence of a passing
animal, gained from a lifetime of hunting, was extraordinary. He
gently glided along the path, sometimes stooping to get a closer
look, pausing now and again to listen, and I found watching him like
observing a dancer in an unhurried but carefully choreographed outdoor ballet. While he walked, he would explain what he was doing
in a soft, purring way, punctuated by the popping sounds that distinguish his language. He was passing on his rich store of knowledge
to two boys from the tribe, ensuring that it didn't die with him. Observing their interactions, I wondered how many thousands of generations had acted out this same scene, the elder teaching the old ways
to the future hunters. I also wondered how much longer this would
go on, with the twenty-first century rapidly encroaching on the
group. Their hunting grounds were in danger of being sold to a safari
outfit from the United Arab Emirates, who wanted to bring well-
heeled men from the Persian Gulf here to hunt, their large-caliber
guns making a mockery of Julius's slender bow and delicate arrows.

Julius is a Hadzabe, among the last remaining members of an
ancient group that has lived in Tanzania for tens of thousands of
years. Their unusual language, with clicks forming a part of many
words, like the Khoisan languages spoken by their hunter-gatherer

cousins in southern Africa, is completely different from that spoken by the nearby Masai cattle herders. During the dry season they live in shelters made from branches and grasses woven together into a basic tentlike structure, and in the wet season they move into a cave shelter closer to Ngorongoro Crater. The location of their dry season camps is dictated by the hunt—when they kill a large animal, the entire group of twenty or so will relocate to that place. This seminomadic hunter-gatherer existence is a perfect adaptation to life on the game-rich African savanna.

Like hunter-gatherers elsewhere, the Hadzabe live with virtually no modern conveniences. Although Julius spends part of the year working as a guide and tracker at a nearby game park in order to make money to help support his family—even hunter-gatherers sometimes need medicines and other modern necessities—most of the time he chooses to live in the traditional Hadzabe way, near Lake Eyasi. This means making nearly everything, the only concessions being a few metal tools: an ax, knives, and small pieces of metal that they hammer into the tips for their arrows. To spend time living with the Hadzabe is to return to an ancient way of life, one where the term "self-sufficient" takes on new meaning. They possess a huge body of knowledge on natural history, including which plants can be eaten and which produce potent poison that can be used to kill large game; how to identify the species, age, and even the sex of the animals from their tracks; how to tell bird species from their calls; and so on. Hadzabe boys learn how to make their own arrows, carefully straightening selected branches by warming them in a fire and bending them with their teeth; then they practice shooting until they are able to hit their targets with lethal accuracy. The Hadzabe spend their evenings telling stories of recent hunts and ancient legends while sitting around a small fire, story lines punctuated with careful intonations, sound effects, and jokes. It is a rich and varied existence, and after several days of living with them I started to feel a distinct calmness, as the worries and clutter of modern life melted away. In an odd way, it felt like returning home after a long absence.

For most of us, living in the way the Hadzabe do is unimagin-

**FIGURE 30: HADZABE BOW AND ARROWS.**

able. It has been so many generations since we last hunted and gathered our food that we no longer have the skills necessary to survive. The closest we typically get to hunting is searching for a parking spot in a crowded urban neighborhood or reaching for the last container of mint chocolate chip ice cream in the grocery store. The difference involves more than just how we obtain our food, though. There's something else about the way people like the Hadzabe live their lives, something that seems to combine so many aspects of what it means to be human, making use of so many different skills. While we may see ourselves as multitaskers, in fact the jobs most of us carry out are remarkably focused. Whether staring at a computer screen all day, sitting in meetings and making conference calls, installing kitchen appliances, playing professional baseball, or driving a truck, we live lives defined by a narrow range of skills. Perhaps this is why spending time with the Hadzabe is so liberating: it forces us to rediscover a treasure trove of long-lost abilities.

Such an admiring view of a group like the Hadzabe runs counter to recent cultural attitudes. One of the great debates of nineteenth-century social theory concerned whether social evolution is progressive. In other words, is there some purpose to our lives, and are the exertions of the present leading us to a brighter future? Early evolutionists represented life as a great chain of being, with single-celled organisms at the bottom of the chain and humans at the top, as the pinnacle of the evolutionary process. Furthermore, even humans were divided into those who had achieved a certain level of material culture ("civilized") and those who in some way lagged behind ("uncivilized"). Lewis Morgan, an influential nineteenth-century social theorist, wrote in his 1877 magnum opus, *Ancient Society,* about three stages of human cultural evolution: savagery, barbarism, and civilization. Savagery was hunting and gathering, the way humanity had lived during its earliest stages of evolution. Barbarism was essentially subsistence agriculture of the type practiced during the Neolithic. Civilization was advanced agriculture, urban settlements, written language, and the rule of law. Morgan's evolutionary approach to human cultures greatly influenced Friedrich Engels, who, of course, had his own idea of where social evolution was headed. The theme of nineteenth-century progressive thinking was that the goal of humanity was to arrive at the form of civilization found in Europe and other "advanced" societies, and that alternative ways of life were primitive and undesirable. In fact, according to Morgan, it was inevitable that cultural evolution would follow such a trajectory, with the advanced societies triumphing over those stuck in an earlier evolutionary rut.

In the case of the Hadzabe, successive waves of agriculturalists and imperial European powers took over their land, forcing them into the small enclave where fewer than two thousand scrape out a meager living today. Their lack of agriculture, land ownership, advanced tools, and money were held up as examples of why such a primitive people needed to be conquered and civilized. Their children were taken away to missionary schools, their names were changed (Julius's real name is !Um!um!ume, with the exclamation mark representing the *tsk* sound you might use when scolding a child), their culture represented as backward and undeserving of preservation, and their land given to other people by the Tanzanian government. With the rise in cultural relativism in twentieth-century anthropological thinking, it was eventually possible to appreciate the variety of human cultures as equally valid, but this shift in attitude was nearly too late for the Hadzabe and other hunter-gatherer groups around the world, for whom the damage was already done. Today, if they still follow a traditional way of life at all, it is as part of a conscious effort, effectively making themselves into living museum pieces as a cultural statement or for tourist dollars.

We have seen elsewhere in this book how and why the cultural progression away from hunting and gathering took place, and the effects it has had on our bodies, our society, and our planet. What I'm focusing on in this final chapter is less quantifiable and more subtle: the effects on our "moral compasses." How do we define morality in today's complex world? Is it possible to learn anything from people like the Hadzabe who live outside the mainstream of modern society in a way that our ancestors did during the many millions of years that our brains and behavioral norms were evolving? Clearly there can be no easy answers to questions like these, as they involve many complex philosophical ideas. However, taking an evolutionary view of how our moral systems came into being is a worthwhile exercise. Many people today would argue that modern society lacks morality. If so, how did we reach this point? One has only to look at the events of September 11, 2001, to understand that something, somewhere is not right.

How we turn our collective moral compass in a positive and sustainable direction and at the same time try to deal with technological advances in fields like genetics and the effects of global warming is one of the great problems of this century. It is also probably the most difficult to solve, if it can indeed be said to be solvable, as there are so many competing interpretations of morality. Philosophers, religious leaders, writers, legal scholars, and politicians have spent thousands of years trying to develop a coherent ethical framework for living one's life, for knowing what is "right" and what is "wrong," and for providing guidance in even the most trivial scenarios (think of newspaper advice columns). To attempt to survey the entire field of thought is not only impossible, it's perhaps even counterproductive. Sometimes it is worth starting with a blank page and trying to reconstruct how we arrived at our present situation—in other words, taking a scientific and historical approach to a very modern problem. So let's begin with a mathematical approach.

## PRISONERS, METAETHICS, AND GREED

Game theory is a branch of applied mathematics developed by the Hungarian-American mathematician John von Neumann in the first half of the twentieth century. In 1944, with the economist Oskar Morgenstern, he published the hugely influential book *Theory of Games and Economic Behavior*. The basic idea is to take social situations and formalize them as "games" that can be analyzed using mathematics. This approach to human behavior has subsequently been widely applied not only in economics but also in other social and biological sciences. In fact, it is one of the most influential developments in twentieth-century mathematics.

One of the classic problems investigated by game theoreticians was developed in the 1950s. Known as the prisoner's dilemma, it provides a framework in which to investigate the development of morality. The scenario is as follows: two suspects are arrested for a crime that they may have committed and are separated at the station house. The police don't have enough evidence to convict, but by keeping them separate during the interrogation they hope to get one prisoner to betray the other. If this happens the betrayer will be released and the betrayed will receive a sentence of ten years in prison. If both betray each other, each will receive a sentence of five years. If, on the other hand, neither betrays the other, each will receive only six months in prison. What should they do, in a rational world?

A careful mathematical analysis reveals that the most rational strategy is for each prisoner to betray the other, since there is a chance that the other person won't betray you and you'll go free. This is known as the *classical* prisoner's dilemma. If, on the other hand, the game is played many times in succession, it is known as the *iterated* prisoner's dilemma. If the prisoners know that the game will be played only a specified number of times, then the best solution is still for each person to betray the other. If, however, the number of rounds is unknown, a new strategy becomes more successful: each player not betraying the other, so that they each spend six months in prison. In the parlance of the game theory field, it becomes evolutionarily stable to cooperate, since you don't know how many times you will be playing again in the future, and if you betray the other prisoner in this round, he or she could betray you in the next.

Obviously, such a simple game is a gross simplification of human social interactions, but it does give us a tool to begin to dissect what we mean by good—in other words, how morality can be defined, and how notions of good and bad might have arisen in the first place. In the case of the prisoners, each knows that there is a chance in any given turn that he or she will be betrayed by the other, and that if this happens the next round could result in retaliation. The fear of retaliation, in effect, keeps one's baser instincts in check. After all, it's better to spend six months in jail than five or ten years, even if this means giving up the chance of going free. Such an approach to the concept of cooperation was outlined— without his knowing about the prisoner's dilemma, of course—by Plato in his *Republic.* In Book II, one of the characters, Glaucon, describes the origins of morality as a compromise between the desire to do injustice—to betray, in prisoner-speak—and the knowledge that injustice might happen to you. This contrasts with Socrates's view that there is something inherently desirable about being just, or moral, even if there is no threat of punishment.

These two opposing views of morality have been debated ever since Plato's day, and no clear winner has emerged. The field of philosophy known as metaethics attempts to answer questions about the origins of morality and what one should do in various situations. Is ethical behavior defined only in reference to a particular society, for instance, or are there universals that would be regarded as moral by everyone? It certainly isn't possible in this chapter to examine the entire history of this quest for moral explanations, but what seems to be clear is that there probably are some universals, and that they likely extend from the sort of conundrum posed by the prisoner's dilemma. In human society, no one is a free agent; our actions always have potential impacts on others. Most laws in modern secular society derive from this premise, and punishments are meted out for people whose actions impinge on others' lives: murder, assault, theft, or behaviors that might increase the likelihood of these happening (e.g., driving at double the speed limit while intoxicated).

Religion, too, has provided guidance on these same ethical questions, attempting to define morality partly in terms of an absence of prohibited behaviors, or sins. In addition to murder, most religions have prohibitions on adultery, greed, and so on. The seven deadly sins defined by the Catholic church are lust, gluttony, greed, sloth, wrath, envy, and pride. Boil them down and they basically fall into four categories: laziness, lust, rage, and wanting more than you have.

The first three, while considered important, clearly weren't emphasized as much as the fourth, which gets four sins devoted to it, including what was considered the most serious, pride. Think about it, though: wanting more than you have is actually central to modern life. Without it we would still be sitting around a campfire somewhere on the savanna, as the saying goes. Gordon Gekko's infamous creed "Greed is good" in the film *Wall Street* could only have been uttered in a society in which it is possible to accumulate excess—in other words, an agricultural society of the sort that we've had for the past 10,000 years. In the world of the hunter-gatherer there can be no excess, since everyone has what he or she needs and accumulating excess would make no sense. Why kill more animals than you can eat when the meat will only spoil? But accumulating surpluses is desirable, even necessary, in a society where resources are limited, since by doing so you assure yourself and your family of access to these resources in the future. Wandering from one minimum-wage job to another, living paycheck to paycheck, is not a lifestyle most of us in the modern would want for ourselves or our children. Essentially, though, this is the life led by hunter-gatherers, and by our ancestors, since they were always reasonably sure that they would be able to find enough of the resources they needed to survive.

Jean-Jacques Rousseau, the eighteenth-century French philosopher and social critic, is perhaps most famous for his depiction of primitive peoples (as well as humanity in the distant past) as living in a state of grace—a view that has come to be encapsulated by the term "noble savage." Rousseau felt that humans are inherently good, and that it is society that has corrupted us. Friedrich Engels espoused essentially the same position, as have other writers. The idea is that if we were only able to return to a state of nature (whatever form that might take), all of our problems would be washed away and there would be no need for law enforcement. The opposite view was voiced by Thomas Hobbes in his description of the Leviathan. According to Hobbes, the "nasty, brutish, and short" lives suffered by savages were mitigated only through the apparatus of the state.

These two opposing views of humanity have been debated for over two centuries, with no clear winner. In the 1960s, encouraged by the influence of the anti-establishment counterculture in intellectual circles, Rousseau's view gained popularity. At the influential "Man the Hunter" anthropology conference held in Chicago in 1966, much of the discussion centered on the lifestyle of hunter-gatherers. This is where Marshall Sahlins coined the phrase "the original affluent society," referring to the fact that hunter-gatherers seemed to have had everything they needed—though not in excess, of course—and worked far less than most Westerners, allowing them the ultimate affluence, that of free time.

Much criticism of Sahlins's view has focused on the relatively small data set on which it was based and the fact that the anthropological observations that led to it were minimal and potentially biased. Today we seem to have turned in the opposite direction, and Rousseau's views are often portrayed as naïve. Anthropologists have attempted to show that "primitive" peoples have at least as much capacity for immorality (for lack of a better term) as modern civilizations. Lawrence Keeley, in his book *War Before Civilization,* presents an impressive amount of data on violence in such primitive societies, finding that warfare is common. He provides a wealth of statistical support for his argument, demonstrating that a large fraction of the people living in such primitive groups die violent deaths as a result of intergroup clashes. The Dani of the New Guinea highlands, for instance, lose one man in three to murder. Even the bloodiest wars of the twentieth century haven't usually killed more than 10 percent of the adult male population of any particular country (Nazi Germany and the Soviet Union in the Second World War being notable exceptions), and for the average American man today, the lifetime chances of being murdered are roughly one in one hundred. Clearly, Keeley argues, primitive peoples are not living in a state of grace.

Keeley's analysis often groups all preindustrial societies together, regardless of their mode of subsistence. When we examine the data on warfare among hunter-gatherers alone, though, his argument becomes a bit more tenuous. As detailed by Keeley in his book, anthropologist Keith Otterbein compiled data on fifty societies around the world and made comparisons on the basis of their way of life. Those engaging in agriculture and animal husbandry had the highest frequencies of warfare, with more than 90 percent engaged in warfare continuously or frequently. Of those societies that "rarely or never" engaged in warfare, most were hunter-gatherers—30 percent of the hunter-gatherer groups studied fell into this category. While the small number of groups surveyed makes the significance of this difference debatable, it is suggestive. The fact that the world's remaining hunter-gatherers are typically marginalized people living in territories that have been reduced by the surrounding agricultural populations, and thus hardly exist in a "natural" state, also complicates the comparison.

One potential insight into the issue of hunter-gatherer morality comes from the Moriori, who live in the Chatham Islands of the South Pacific and provide an interesting case study of the effects of subsistence mode on violence. An offshoot of the better-known Maori living farther west, in New Zealand (a Polynesian people renowned for their violent intergroup warfare), the Moriori are thought to have migrated to their homeland around the year 1500. The Chathams were not suited to the typical Polynesian crops grown by the agricultural Maori, however, so the Moriori had to return to hunting and gathering, living off the abundant coastal resources in the islands. Their culture, presumably originally as warlike as that of the Maori, changed over time. They became dedicated pacifists, settling arguments with ritualized battles and discussion, since warfare made no sense in their new homeland and with their way of life. The islands were annexed by the British in 1791, and in 1835 nearly a thousand Maori carried there by British ships massacred the two thousand Moriori estimated to live there. The Moriori were essentially extinguished as a people, and within a generation only around one hundred survived, conquered by their far more violent agricultural cousins, their demise helped along by European diseases.

Perhaps the Moriori are a fluke, a local oddity. But as we saw in

Chapter 2, societies typically engage in warfare when resources are limited and there is a high degree of competition. For hunter-gatherers this is not typically the case, unless they are living in an unusual situation, such as the inhabitants of Jebel Sahaba, in Nubia, who relied on fishing in an increasingly arid environment. During most of human history resources were not limited, and the human population expanded throughout the world. When we developed agriculture, of course, the increase in population density meant that the days of easy excess were gone; if you wanted or needed more food, a great deal of investment was required to grow it. Tending fields, maintaining irrigation systems, building permanent settlements—all of it meant that there was too much at stake for anyone to wander off and find another place to live and, therefore, that one's investment was worth fighting for. While we likely entered into our new lifestyle in response to climatic challenges that made hunting and gathering more difficult, once we had done so we created a new set of competitive challenges.

I'm not advocating a return to a hunter-gatherer lifestyle, of course—merely pointing out that we can learn something about the state of modern society from those ancestors. Some people, wary of the constant struggle to work and accumulate inherent in modern life, have chosen to live outside the system, or even to actively rebel. The "turn on, tune in, drop out" attitude espoused by Timothy Leary in the 1960s is simply a recent example of such a rebellion. The Luddites of the early 1800s, English cottage weavers who destroyed factories in order to save their jobs and their traditional way of life, were early opponents of modernization and the industrial era. Less violent was the nineteenth-century Romantic movement, steeped in Rousseau, its proponents eager to return to nature and simpler times. Marxists and Leninists embraced modernity but not the capitalist economic system that had spawned it. The Amish and other religious groups who choose to live a life peripheral to the mainstream of Western society are consciously rebelling against what they see as the corrupting influences of technology and other trappings of modernity. Anti-globalization protestors decry the loss of local industries to an American-led cultural juggernaut, and the Slow Food movement urges us to eat simple, local foods free of the corrupting effects of McDonald's and KFC. All of these movements, regardless of their particular focus, have been formed in response to what are seen by their members as the dangers of modernity.

Over the past half century another anti-progress trend has been spawned, one more widespread and potentially dangerous than the more limited movements of the past: fundamentalism. Coalescing in the mid–twentieth century in both the Islamic and the Christian faiths, fundamentalism has increasingly dominated political debates around the world. Born of desperation and anger, and driven forward by charismatic leaders, fundamentalist views provide a focus for people who feel left out of the modern world, offering an alternative vision of how life should be lived. Its rise, and what it might mean for the future, is where we're headed next.

## FUNDAMENTALISM

In the 1950s, Egypt was going through an unprecedented transformation. Having been part of the British Empire for a short period during and after the First World War, and a virtual client state of Britain during the three decades after it was declared "independent" in 1922, it finally took the political future into its own hands in 1952 when a group of army officers led by Gamal Abdel Nasser overthrew the government in a coup. Nasser was inspired by what he saw as the success of the Soviet Union, and he immediately set about remaking Egypt on a socialist model. Part of his plan included the secularization of Egyptian society. Though the new regime was still technically Muslim, Nasser and his government believed strongly in the separation of mullah and state. Islam was something to be sidelined from the primary goal of development along secular, socialist lines.

All of this was happening at a time when Arabs felt weakened by the creation of Israel in 1948. A Jewish state occupying one of

Islam's holiest cities was something that many could not forgive, and throughout the Arab world it was known as "the catastrophe." The Arab-Israeli War, which followed Israel's declaration of independence from British rule, was won decisively by the Israelis, further deflating Islam's collective ego. It was against this backdrop that some Muslims began to explore alternatives.

The Muslim Brotherhood, an Islamic cultural group, was founded in the 1920s as Egypt's fortunes were in decline. By the late 1940s it had millions of members and was a powerful combination of trade union, cultural organization, and political party whose influence reached into every aspect of Egyptian life. The Brotherhood espoused a philosophy that demanded a strict, all-encompassing adherence to the tenets of Islam—not simply practicing its tenets, like praying five times daily or undertaking the hajj, but living Islam. Its insistence on implementing *sharia,* or religious law, earned it many enemies in the Egyptian government. While not founded on violent principles, but as an educational organization meant to spread the word about Islam, it became increasingly militant over the years. In 1949 one of its members assassinated the Egyptian prime minister, which led to a retaliatory assassination of the Brotherhood's founder, Hassan al-Banna, by government forces.

Sayyid Qutb (pronounced SIGH-eed KOO-tub) was an Egyptian author and literary critic who became a member of the Brotherhood in 1953, at the age of forty-seven. By this point the organization had been outlawed by the Egyptian government as part of Nasser's rush to secularize the country. Qutb was among many Brotherhood members arrested in the fifties, and he was sentenced to fifteen years of hard labor in 1954. His time in prison, understandably, made a strong impression on him, but instead of rebelling openly or giving up and becoming a broken man, he fought back using that most powerful of prisoners' weapons: the pen. During his years in jail he wrote two books that would turn out to be hugely influential in Islam: a commentary on the Koran and *Ma'alim fi al-Tariq* (*Milestones*). While the former was a scholarly, multivolume treatise, the latter was a manifesto.

In *Milestones,* published in 1964, Qutb argued that it was time for Islam to reclaim its supremacy as a world religion. The reason for its decline, he maintained, was that Muslims had surrendered to *jahiliyyah,* or ignorance of God; in other words, secularism had led to its decline. The only way to reverse this decline, he said, was to surrender completely to an Islamic way of life: *sharia* and a rejection of the state of *jahiliyyah.* The latter would be accomplished in part through *jihad* (struggle), a holy war against corrupt secularism.

Qutb's focus on *jihad* was new, as were the methods he advocated in order to achieve the liberation of Islam. Since the time of Muhammad, Muslims had been engaged in struggle in the name of religion; this was the reason for the spectacular territorial gains of Muhammad's followers in the Middle East and North Africa during his lifetime and immediately following his death. It was never seen as a primary goal, or tenet, of the religion, however. Qutb reinterpreted Islam within the scope of the modern world, giving it a new focus. In the same way that Lenin's interpretation of Marxism was used to justify the Russian Revolution, Qutb hoped that his work would lead to an Islamic revolution.

In her book on the history of fundamentalism, *The Battle for God,* Karen Armstrong outlines the debate between two opposing realms of thought that have dominated humanity's worldview, probably since the origin of our species tens of thousands of years ago. *Mythos* is a mystical way of viewing the world, one preoccupied with received meanings about significant events. *Logos*—the word provides the Greek root for the word "logic"—is the realm of rationalism, science, and Enlightenment thought. *Mythos* is about accepting the spiritual aspects of the world, while *logos* is concerned with questioning and understanding. For thousands of years, human societies have incorporated aspects of both, but in the past few centuries *logos* has come to the fore. It underpins scientific thought, has provided us with the wonders of modern technology, and has led to unprecedented levels of wealth. In its ascendancy it has also, many would argue, led to the destruction of the old certainties that so many relied on to give their lives meaning. It is this latter view that Qutb was espousing, but with a novel twist: he wanted to apply a *logos* approach to achieving *mythos* ends. In other words, the decline of Islam and the rise of global *jahiliyyah* was a problem to be solved, and the solution lay in *jihad.*

At the same time that Qutb's views were receiving widespread attention in the Islamic world, many Americans were feeling the loss of *mythos* in their own lives. The social upheavals of the 1960s, with their rejection of traditional Christian family values, were a shock to many. The rise of Christian fundamentalism in the United States was, even more than the spread of Qutb's scholarly manifesto, an application of *logos* thinking to a *mythos* problem. At the forefront of the movement was Jerry Falwell, a southern preacher who founded his congregation—the Thomas Road Baptist Church—in Lynchburg, Virginia, in 1956. Using the power of the new mass communications media of radio and television, he soon became known to millions around the country. According to Karen Armstrong, in the 1960s and 1970s, 40 percent of American households tuned in to his broadcasts. The *logos* of modern technology was being used in a novel way to mobilize a conservative social movement, its goal being to return religion—*mythos*—to its rightful place at the center of society.

In the 1970s, Pat Robertson and Jim and Tammy Faye Bakker would follow Falwell's lead. Watergate, a demoralizing withdrawal from Vietnam, the oil crisis of the 1970s, and the Iranian hostage crisis would all combine to challenge America's sense that it was on the right track. By the time of the 1980 presidential election, many people sensed that the time had come for more religion in the political process. Falwell founded the Moral Majority in 1979, and Ronald Reagan successfully courted that group to win the election with a landslide 489 electoral college votes to Jimmy Carter's 49. From that point onward, the Republican Party—traditionally the party of the industrial Northeast and big business—would be realigned with the rural South, family values, and Falwell's Moral Majority. Religion would take its place on the American political stage in a way it never had before.

The 1980s would see the concepts of *jihad* and "family values" consolidated and formalized in the Islamic and American spheres of influence. The fundamentalists on the Islamic side shifted to battle mode, inspired in part by Ayatollah Khomeini and the 1979 Islamic Revolution in Iran. While various Arab organizations had always had radical factions that carried out political assassinations—including the Muslim Brotherhood—they'd typically fought for political power or territorial gains, and their targets had often been highly visible people in the opposing camp. The new interpretation of *jihad* was much broader. The group Islamic Jihad arrived on the scene in 1983 when it claimed responsibility for the bombing of the U.S. Marine Corps barracks and the American embassy in Beirut, and many elements of the organization were incorporated into the Lebanese group Hezbollah later in the decade. Hamas, a radical Palestinian spin-off of the Muslim Brotherhood intent on establishing a separate Palestinian state, was founded in 1987. Al-Qaeda, certainly the most notorious Islamic terrorist organization of the past decade, was founded in 1988. It emerged in the wake of the covert, CIA-funded war in Afghanistan with the aspirations of many *mujahideen* to continue the battle against non-Muslims elsewhere in the Islamic world, notably Israel.

Al-Qaeda in particular has taken Sayyid Qutb's writings to heart, and much of Osama bin Laden's rhetoric is lifted directly from his writings. The organization's recruiting materials espouse radical interpretations of Islam that bear no resemblance to the religion practiced by the vast majority of the world's Muslims. In fact, this is used by the organization to justify killing them in the name of *jihad*: Muslims who do not practice the extreme form of Islam championed by al-Qaeda are considered by the organization to be *takfir*—apostate sinners who are, in effect, denouncing their religion. We can trace this belief directly to Qutb, who said that it was permissible in the name of *jihad* to kill such people.

This history helps to explain how we ended up in the frightening position in which we find ourselves today. Members of the American religious right killing doctors at abortion clinics; al-

Qaeda members using bombs and hijacked planes to kill thousands in a global war against Western secularism; Aum Shinrikyo followers killing a dozen people and injuring hundreds more in the Tokyo subway—all seem like the inhuman acts of crazy zealots, unable to separate right from wrong, but this is a crude oversimplification. Any sort of mass murder is possible only because the human faces of the victims have been dissociated from the action of killing; unlike previous movements that have advocated violent solutions to social problems, however, such as the PLO or the IRA, with their territorial desires, today's violent fundamentalist movements claim to be doing God's will, giving them a sense of higher purpose. Success means not simply achieving proximal goals such as political autonomy, for instance, but changing the world order in the name of God. Crazy though it may appear from the outside, it is not insanity that drives terrorism in the fundamentalist world; rather, it is the God-given certainty that what one is doing is morally just.

Of course, most fundamentalists will never commit acts of terrorism. Nevertheless, their decisions on a daily basis help to forge an opposition to liberal secularism. The battle over evolution in the United States is a good example. When Charles Darwin first proposed his theory of evolution by natural selection in 1859, it met with widespread opposition from the religious establishment. Bishop Samuel Wilberforce, who famously debated evolution with Darwin's ardent supporter T. H. Huxley in 1860, parodied the notion that humans could be descended from apes, but such views were ultimately to fade into the minority. By the early twentieth century, Darwinian evolution was widely accepted, in part because of its misappropriation by proponents of "social Darwinism" as a way to justify the hardscrabble economic competition of the Victorian era. While the famous Scopes "Monkey Trial," in 1926, briefly brought anti-evolutionist thinking back into public view, ultimately the Great Depression, the Second World War, and the growing prominence of science during the Cold War would squelch much of this fundamentalist backlash. A 1968 Supreme Court decision overturning an Arkansas law banning the teaching of evolution in schools effectively ended the debate for the next decade.

In the 1980s, flush with Ronald Reagan's victory in the presidential election and the concomitant increase in the prominence of the Moral Majority, anti-evolutionist thinking began to claw back support it had lost during the secular mid-twentieth century. A study published in 2006, examining public acceptance of evolution around the world, found that around 80 percent of the people in major European countries and Japan accepted evolution. In the United States the number was only 40 percent, and among the thirty-four countries surveyed only Turkey ranked lower, at 25 percent. The poll figure in the United States actually represented a decline from 45 percent in 1985, and between 1985 and 2006 the number of people who were "unsure" about evolution increased from 7 percent to 21 percent. One of the study's coauthors, Jon Miller of Michigan State University, noted that "American Protestantism is more fundamentalist than anybody except perhaps the Islamic fundamentalist."

Clearly, the past few decades have seen an unprecedented rise in fundamentalist thinking as people have felt increasingly left out of the secular future. According to Stephen Ulph of the Jamestown Foundation, a think tank focused on terrorism, at least 60 percent of the material on Islamic jihadist websites concerns ideological or cultural questions rather than pragmatic issues. He noted in an interview in the *Economist* that jihadists are fighting less a war against the West than "a civil war for the minds of Muslim youth." Similarly, American fundamentalists view their struggle in terms of a battle with the forces of secular humanism for the hearts and minds of America—according to the Moral Majority Coalition website, they plan to "mobilize people of faith to reclaim this great country." Both movements, while rooted in their desire to return to what they see as a golden age of religious *mythos,* are making ever-greater use of the methods of *logos* in their mobilization. In their effort to return to a past they imagine to be more pure, they are using the technologies of the present to reach out to more and more people.

A few hundred years ago, the world might as well have been composed of separate planets, each one with its own culture. In some (e.g., South America or Africa) *mythos* dominated, with significant emphasis on religion and tradition as the bedrock of society. Others (those in western Europe, for instance) were rushing toward a *logos* establishment. What happened over the succeeding centuries is that these previously separate planets were thrown into contact, often violently, in a way that created the modern *logos*-dominated world. Those cultures that looked to *mythos* for guidance were either marginalized or conquered, their people incorporated into the secular. What we are witnessing with the rise of fundamentalism in the past half century is a process whereby some of the people who sense the loss of *mythos* are trying to regain it. One of the dominant features of fundamentalist teachings, regardless of the religious sources underpinning them, is that of separatism—a desire to build an alternative culture outside the mainstream of the modern world.

The world, it seems, has always been Balkanized in this way. It allows us to make sense of the mass of humanity we jostle with in modern society. The American melting pot has always been curdled, with hyphenated Americans (African-Americans, Italian-Americans, and so on) the norm. The world today is far more mobile than it has ever been, and cultures are mixing to a far greater extent than they have before. Rewind the clock a few centuries, back before the dawn of the European Age of Exploration, and it would have been a different place. There were perhaps fifteen thousand languages spoken in the year 1500, as opposed to today's six thousand—and those six thousand will probably be reduced to fewer than three thousand by the end of this century. This is one way to chart the march of globalization, as previously isolated groups all over the world join the global melting pot at an ever-increasing rate.

This desire to find a small community within the dizzying demographic cacophony of the modern world is nothing new. Whether inspired by fundamentalism or not, it is something we do all the time, even in the most modern places.

## FACEBOOK FORAGERS

One of the things I do for amusement on the Internet is to share information about myself with a small group of friends on Facebook.
The status updates are always an opportunity to be clever, funny, or
offbeat while letting your friends know what you are up to; along
with uploaded photographs and shared YouTube videos, the site
gives you a chance to carve out an online presence in a forum that's
substantially more intimate than a blog. I'm hardly alone in my
Facebook activities, either—as of early 2010 it boasted more than
400 million members in more than 180 countries, and it was
growing at a rate of around 10 million members a month. Begun
in 2003 by Mark Zuckerberg, then a Harvard undergraduate, as a
way for his fellow classmates to stay in touch with each other, its
growth has been nothing short of phenomenal. If Facebook were a
country, it would be the world's third largest, smaller only than
China and India. But what an unusual country it is.

Facebook reveals details of what you post only to your friends—
people you know and are willing to share such information with.
Unlike a real country of hundreds of millions of people, you can't
simply walk down a street in Facebook and bump into random
strangers. Actually, you could, but such random introductions
would likely be met with a brush-off where the person approached
for "friendship" clicked the "Ignore" button on the screen. There is
a whole mini industry built around Facebook etiquette, attempting to sift through what is and isn't advisable to share with your
community. Embarrassing photos from last weekend's party? Probably not a good idea, especially if your friends include work colleagues. Membership in a radical organization? Perhaps better left
unmentioned. It is this combination of publishing the details of
one's life to a group of people while retaining control over what is
revealed that makes Facebook and other social networking sites so
popular.

Despite its position at the forefront of the Web 2.0 revolution, the powerful servers that provide its infrastructure, and the esoteric topics that can be found in its fan clubs and groups (want to let everyone know how much you love seaweed?), what is so striking is how much like the real world Facebook is. The most popular fan sites are for Barack Obama (who famously used Facebook to build support during his election campaign), Coca-Cola, Nutella hazelnut spread, pizza, and the television character Homer Simpson. Pop stars, other foods, and film and sports personalities round out the top twenty. In their Facebook world, users are consciously recreating the aspects they love about the real world, populating their virtual communities with people and items from their everyday lives.

Not so obvious, but still present, is something that creates a link between the brave new world of the Internet and our distant past in hunter-gatherer groups. The average number of Facebook friends people have is around 130. This is close to the number we learned about in Chapter 4—Dunbar's average number of close acquaintances, predicted by the size of the human neocortex. It seems that even in a virtual world where we could have thousands or millions of "friends," we don't want to tell that many people about our fondness for Chinese food or share pictures of our children with them. And perhaps the deep connection between our ancient hunter-gatherer brains and the worlds we create online can show us a path through the ideological thicket, pointing a way toward a new *mythos* in the twenty-first century.

## WANT LESS

One evening, back at camp, I asked Julius how the Hadzabe dealt with their dead—burial or, perhaps, cremation? He smiled and shook his head. "No," he explained, "we don't bury the bodies or burn them. We leave them out in the open in a sacred place, near a baobab tree, and let the animals eat them. Usually it is the hyena. There is a small ceremony performed by the elders, and we leave food for the passage to the afterlife. We return a few months later to find the bones, and then hide them in the bush. The land we live in is full of the remains of our ancestors, and it connects us to them."

The simplicity of his answer is what struck me most. The entirety of the Hadzabes' land is a burial ground, with no separate place for the dead to spend their afterlives. They continue on in the landscape after they are dead, and this is part of the reason that the Hadzabe feel such a close connection to their territory. Land, to the Hadzabe, is not just about something that provides subsistence, though of course that is important—it's also a tangible connection with their ancestors. The land and the people are inseparable, and this connection is the essence of their *mythos.* To attempt to separate them or to exploit the land in some way would be an insult to the entire Hadzabe worldview. It was and is unthinkable.

Many traditional cultures around the world feel such a connection to the land where they live, but perhaps none more so than the hunter-gatherer. The agriculturalist relies on favorable weather to grow his crops, but the *mythos* of agriculture is ultimately one of production and extraction: if I plant the seeds, tend them, protect them from pests, and harvest them, then I will be able to feed my family. Through my labor I change the environment into something I can control, and that allows me to grow enough food to survive. If the land ceases to be productive, I will have to move on to find new land. For agriculturalists, the land itself is simply a conduit, a way to turn seeds into something useful. Slash-and-burn agriculture is simply the most egregious example of this approach to land use.

The problem with such a *mythos,* as any environmentalist will tell you, is that it is simply not sustainable in the long run. Ultimately, the era of Manifest Destiny must come to an end when there is nowhere else to move. The seeds that set in motion the Neolithic Revolution have bloomed into a wholesale rush to extract and exploit most of the earth's resources, largely because it was possible to do so and it seemed that their supply was infinite. But as we enter the twenty-first century, this old *mythos* is becoming increasingly untenable, as we saw in the previous chapter. As resources become more limited, that scarcity challenges the old models of development and progress.

Similarly, the lurch toward fundamentalism in the Islamic world and in the United States has been largely a reaction to the increasing loss of *mythos*. The loss of cultural traditions leaves many people feeling adrift and exploited, human mirrors of our ravaged ecosystems. *Logos* is great at explaining *how* things happen, but it is less helpful in providing an answer to *why*. Humans need such explanations to make sense of the world, and if we don't know about the science—or don't buy into that worldview—we will invoke religious or mythological explanations. Does the sun rise every day because the earth rotates on its axis or because the gods are smiling on us and want to provide its life-giving rays? It all depends on what culture you grew up in.

I am not advocating conversion to fundamentalism, any more than I am urging us to abandon our plows and return to hunting and gathering. But I do think that the current cultural conflict should cause both sides to reexamine their underlying tenets. The modern, secular West should ask what it is that fuels the flames of fundamentalism, just as the advocates of *jihad* should ask why such a war is justified. Ultimately, fundamentalism can exist only in opposition to something else; it is a protest movement. If there were nothing to protest, it would lose its *raison d'être*.

What can we do to encourage such a consilience? The Scottish philosopher David Hume pointed out that *is* is not the same as *ought*—simply because you *can* do something doesn't mean you *should*. But exactly the opposite seems to be the case today: if we *can* do something, that seems to provide a *justification* for doing it. And as we learn to do more things, we want even more—a vicious cycle thriving off the illusion that resources are unlimited. This process started 60,000 years ago as our species expanded from its African homeland to populate the world, and it accelerated abruptly after the Neolithic Revolution. With agriculture, as we have seen, came the power to create far larger problems than we could have even dreamed of as hunter-gatherers, and the driving force behind most of them was greed. While we can never go back to the preagricultural era, we can perhaps take as a moral guide the *mythos* of the world's remaining hunter-gatherers: we can learn to *want less.*

It is only by wanting less that we will be able to come to terms with the challenges of climate change. Wanting less will also give us a better insight into how best to apply powerful technologies such as genetic engineering, by forcing us to accept that there are limits to human perfectibility. And, by making a bit of room for *mythos* at the cultural table, it will probably mitigate some of the extreme forms of fundamentalism. After all, it is the "crass materialism" and "cultural imperialism" of the West that is cited by so many Islamic fundamentalists as a reason for *jihad.* Sayyid Qutb developed his hatred for the modern, secular Western lifestyle after living in the United States; he wrote disparagingly of its "utilitarian" culture, "dominated by . . . materialism." Reducing the importance of such materialism wouldn't put Islamic fundamentalism out of business, of course, but one of its major justifications would be gone.

Continued expansion at the rate we have seen in the past few centuries simply isn't tenable. We are currently using far more of the earth's natural resources than will be sustainable in the long run. Even technological innovation won't create unlimited new supplies of fresh water; nor will it allow everyone to live American lifestyles. Knowing when to say no is sometimes more important than discovering a new way to say yes. Perhaps one of the few positive things to come out of the severe global recession we are experiencing as I write this is that materialism has become less important than it was in the past and the late-twentieth-century substitution of money for *mythos* has lost some of its luster.

Writers such as Bill McKibben and James Howard Kunstler have tapped into the new sense that we need to downsize our lifestyles. McKibben, alarmed at the proliferation of "transhumanist" technologies like genetic engineering that threaten to change the essence of what it means to be human, asks, "Must we ever grow in reach and power? Or can we, should we, ever say 'Enough'?" Kunstler, a critic of American suburbia and the author of *The Geography of Nowhere,* wrote in an op-ed piece in the *Washington Post,* "No combination of solar, wind and nuclear power, ethanol, biodiesel, tar sands and used French-fry oil will allow us to power Wal-Mart, Disney World and the interstate highway system—or even a fraction of these things—in the future. We have to make other arrangements." These other arrangements, for Kunstler, involve changing the relationship we have with where we live and closing the door on suburban sprawl. In essence, he's telling us we need to want less: less commuting, smaller houses, more energy-efficient forms of transportation, food that is more cottage than industry.

Designers and architects are also embracing a new *ethos,* trying to make more intelligent use of their materials. The architect William McDonough writes of a new "cradle-to-cradle" approach to design, where the goal is to rethink the old models of manufacturing and building. He asks in his book *Cradle to Cradle,* "What would have happened . . . if the Industrial Revolution had taken place in societies that emphasize the community over the individual, and where people believed not in a cradle-to-grave life cycle, but in reincarnation?" An interesting question, and one that may very well be answered as we search for a new postreligious *mythos* in the secular world. It's a question into which Julius might have some insight.

We are at a critical juncture, a time unlike any other in the history of our species, when our culture threatens to destroy the essence of what it means to be human. It is vital that we take lessons from our past in order to better know ourselves, and to guess at where we *should* go tomorrow. We are in control of our own destinies, perhaps the first generation ever to have such power, but how will we know what to do? My suggestion is that some of these answers should come from Lake Eyasi, as well as Washington, Brussels, and Kyoto.

On January 22, 2003, NASA received its last communication

from the *Pioneer* 10 spacecraft. *Pioneer* had been launched in 1972 as the first object made by humans intended to leave the solar system. When the communication was received, *Pioneer* was over seven billion miles from earth, far beyond the orbit of Pluto, on a journey that would take it to the red star Aldebaran over the next two million years. It is likely that by the time it is found by extraterrestrial life-forms (assuming they're out there), the earth will have been absorbed by an expanding sun and the world as we know it today will be but a distant memory.

Attached to its outer cover, *Pioneer* carries a plaque as a message to other intelligent life. In addition to explaining the location of our planet in the galaxy, it shows a drawing of a man and a woman. They are naked, and the man has his hand raised in a friendly gesture of welcome. Critically, there is no writing on the plaque, no mention of countries or politicians, no descriptions of religions or material wealth. As the first "hello" from our species to life beyond our planet, we chose to emphasize our biological essence.

*Pioneer* can be seen as a metaphor for the future. Its last feeble radio signal, traveling at the speed of light, took nearly twelve hours to reach earth. This means that when we receive a signal from *Pioneer* in our present, it is already twelve hours ahead of us, speeding alone into the unknown depths of space. In effect, it has already begun to leave our present behind. By the time it reaches the Aldebaran system, the length of time that will have elapsed since its launch will be similar to that separating us from our ancestor *Homo erectus.* Human culture will have changed enormously, but we—assuming we are still around—will still be defined by our biology. There is a lesson in this: at the present critical point in human history, where we have the tools to begin to solve some of the problems set in motion by the Neolithic Revolution, saving ourselves will mean accepting human nature, not suppressing it. It will mean reassessing our cultural emphasis on expansion, acquisition, and perfectibility. It will mean learning from peoples that retain a link back to the way we lived for virtually our entire evolutionary history. And it might allow us to stick around for the next two million years.

# Acknowledgments

Sincere thanks to Jonathan Pritchard of the University of Chicago for discussing his research on the effects of selection on the human genome; to Wolfgang Koppe, Grethe Rosenlund, Alex Obach, and Tor Andre Giskegjerde for explaining their work at the Marine Harvest research center in Stavanger, Norway; and to the fishermen of Kerkennah Islands for showing me their traditional fishing methods. Thanks also to Johann Feilacher and Nina Katschnig of the Haus der Kunstler in Maria Gugging for explaining the fascinating work they are doing with the resident artists, and to the artists themselves for letting me visit their home and see their incredible art. Jayson and Michelle Whitaker kindly hosted me for a long afternoon at their home in Derbyshire and patiently told me the amazing story of their son Charlie's long medical odyssey. Thanks to the people of Tuvalu for sharing their island paradise and patiently letting me barrage them with questions; thanks also to Gilliane Le Gaillac and Christopher Horner for explaining their work in Tuvalu and introducing me to several of their friends there, to Semese Alefaio for a wonderful day on his family's *motu* and the surrounding reef, and to Enate Evi of the Tuvalu Environment Department for taking a meeting with me on fairly short notice. Finally, heartfelt thanks to my friend Julius Indaaya !Um!um!ume for allowing me to spend time with him and his Hadzabe tribe on several occasions, and for his ongoing efforts to preserve his people's ancient way of life—keep the faith.

Only the author's name appears on the cover of a book, but of course it takes a team to write one. I want to thank my editor at Random House, Susanna Porter, for believing in this project over the course of the several years it took to deliver, and for patiently reading and commenting on drafts as it metamorphosed into a book. Her deft touch is visible on nearly every page. Thanks also to my agent, Clare Alexander, for her trust in my ability as a writer and for making this project a reality. Thanks to Stefan McGrath and Will Goodlad at Penguin for their continued interest in the book and their supportive comments throughout my writing career—the fact that you're reading this right now is ultimately due to Stefan's faith in my work more than a decade ago. Justin Morrill of the M Factory did a terrific job on the line illustrations in the book—they really help the material come to life. Finally, thanks to my wife, Pam, who has put up with my obsessive discussions of early farming techniques and disease statistics, and my all-too-frequent absences. I adore you.

# Sources and Further Reading

## CHAPTER 1: MYSTERY IN THE MAP

Jonathan Pritchard and colleagues' paper on selection in the human genome was published in *PLoS Biology* 4:e72 (March 2006; available online at www.plosbiology.org). While I was writing this book, two other papers were published on the topic: Sabeti et al., *Nature* 449:913–18 (2007); Hawks et al., *Proceedings of the National Academy of Sciences USA* 104:20753–58 (2007). In addition, *The 10,000 Year Explosion: How Civilization Accelerated Human Evolution,* by Gregory Cochran and Henry Harpending (Basic Books, New York, 2009), describes the genetic evidence for recent selection and the authors' interpretation of the patterns.

Lawrence Angel's paper on the skeletal evidence for stress as eastern Mediterranean populations adopted agriculture appeared on pages 51–73 of *Paleopathology at the Origins of Agriculture* (Academic Press, Orlando, 1984), edited by Mark Nathan Cohen and George J. Armelagos.

## CHAPTER 2: GROWING A NEW CULTURE

Carlos Duarte and his colleagues describe the recent rise of aquaculture in *Science* 316:382–83 (April 2007). The collapse of global fisheries is described by Boris Worm and his colleagues in *Science* 314:787–90 (2006). Gordon Childe was the author of many books on archaeology and the Neolithic; perhaps the best known are *New Light on the Most Ancient East* (Routledge and Kegan Paul Ltd., London, 1952) and *Man Makes Himself* (Watts & Co., London, 1956). The climate maps in Figure 7 were adapted from the much more detailed ones in Petit et al., *Episodes* 23:230–46 (2000).

The climatic fluctuations at the end of the last ice age are explained in Brian Fagan's wonderful book *The Long Summer: How Climate Changed*

*Civilization* (Basic Books, New York, 2004). The imperative to produce more food is discussed in Mark Nathan Cohen's *The Food Crisis in Prehistory* (Yale University Press, New Haven, 1977). American agronomist and anthropologist Jack Harlan carried out the experiments showing that three weeks of grain gathering can yield enough food to support a family for a year. Dani Nadel's work on early grain gathering around the Sea of Galilee is described in *Science* 316:1830–35 (2007). The megafauna extinction patterns shown in Figure 8 were modified from Figure 17.8 in Paul Martin's contribution to *Quaternary Extinctions: A Prehistoric Revolution,* edited by Paul S. Martin and Richard G. Klein. © 1984 The University of Arizona Board of Regents, reprinted by permission of the University of Arizona Press. The subject of humans and ancient extinction events is rather controversial within the archaeological and paleoecology communities; a good short review can be found in *Science* 300:885 (2003). A discussion of strontium and the Neolithic diet can be found on page 371 of *The Cambridge Encyclopedia of Human Evolution* (Cambridge University Press, Cambridge, U.K., 1992), edited by Steve Jones, Robert Martin, and David Pilbeam.

Nikolai Vavilov authored many scientific papers and several books on the origins of crop plants. A good summary can be found in "The Origin, Variation, Immunity and Breeding of Cultivated Plants," published in *Chronica Botanica,* Volume 13, 1949. A monumental overview of the origins of cultivated foods is the two-volume *Cambridge World History of Food* (Cambridge University Press, Cambridge, U.K., 2000), edited by Kenneth F. Kiple and Kriemhild Conee Ornelas. Daniel Zohary and Maria Hopf's *Domestication of Plants in the Old World* (Oxford University Press, Oxford, 2000) is a good review of the subject. Alan Davidson's fascinating and entertaining *Oxford Companion to Food* (Oxford University Press, Oxford, 1999) is another great reference work.

Figure 9 is modified from a figure in N. J. van der Merwe's paper in *American Scientist* 70:596–606 (1982), reprinted in *The Cambridge Encyclopedia of Human Evolution.* Zhijun Zhao's work on early rice domestication is presented in *Geoarchaeology* 15:203–22 (2000). Andrew Moore and his colleagues' work at Abu Hureyra is described in the book *Village on the Euphrates* (Oxford University Press, Oxford, 2000). Susumu Ohno's work on gene duplication is presented in his book *Evolution by Gene Duplication* (Springer-Verlag, Berlin, 1970). The study of the genes involved in domestication was published by Jaenicke-Després and colleagues in *Science*

302:1206–08 (2003). A thorough—if slightly dated—review of many aspects of hunter-gatherer societies is presented in *Man the Hunter* (Aldine, New York, 1968), the proceedings of a conference held at the University of Chicago in April 1966. Another good reference is *The Cambridge Encyclopedia of Hunter-Gatherers*, edited by Richard B. Lee and Richard Daly (Cambridge University Press, Cambridge, U.K., 1999).

James Mellaart's work at Çatalhöyük is presented in his book *Çatal Hüyük: A Neolithic Town in Anatolia* (McGraw-Hill, New York, 1967), though some of the interpretations of his finds have changed since the book was published. Neolithic Venus figures and the accompanying "goddess cult" are described in Marija Gimbutas's book *The Living Goddesses* (University of California Press, Berkeley, 1999).

## CHAPTER 3: DISEASED

The annual report on obesity in the United States is compiled by the Trust for America's Health; its most recent report can be found online at healthyamericans.org. Household income data used in Figure 14 appears in the U.S. Census Bureau publication *Income, Earnings, and Poverty Data* from the 2007 *American Community Survey*, published in 2008. Eric Schlosser's *Fast Food Nation: The Dark Side of the All-American Meal* (Harper Perennial, New York, 2002), charting the rise of fast-food culture in the United States, is a modern classic, as is Michael Pollan's *The Omnivore's Dilemma: A Natural History of Four Meals* (Penguin, New York, 2007).

James Neel's work is described in his autobiography, *Physician to the Gene Pool: Genetic Lessons and Other Stories* (John Wiley & Sons, New York, 1994). The data on diabetes in Samoa is taken from a study by Tsai and colleagues published in the *American Journal of Human Genetics* 69:1236–44 (2001). The Pima Indian diabetes data is from Schulz et al., *Diabetes Care* 29:1866–1971 (2006). The story of the spread of SARS is readily available at several news sites on the Internet. William McNeill's book *Plagues and Peoples* (Anchor, New York, 1976) has had a huge influence on many subsequent works, including Jared Diamond's *Guns, Germs, and Steel* (W. W. Norton, New York, 1997).

The statistics on malaria were taken from the U.S. Centers for Disease Control (CDC) and the World Health Organization. The recent satellite analysis of Angkor by Damien Evans and colleagues, published in *Pro-*

*ceedings of the National Academy of Science USA* 104:14277–82 (2007), has revealed it to be far larger than previously thought—more than 380 square miles in area. Jacques Verdrager's theory on the abandonment of Angkor was published in the journal *Médecine tropicale: revue du Corps de Santé colonial,* 52:377–84 (1992). Deirdre Joy and colleagues' work on the ancient origin of *falciparum* malaria was published in *Science* 300:318–21 (2003). Sarah Tishkoff's work on *G6PD* and recent malarial selection was published in *Science* 293:455–62 (2001).

The evidence for Neolithic dental work at Mehrgarh was published by Roberto Macchiarelli and his colleagues in *Nature* 440:755–56 (2006). Clark Spencer Larsen's work on cavities in North American populations during the agricultural transition is reviewed in his book *Bioarchaeology: Interpreting Behavior from the Human Skeleton* (Cambridge University Press, Cambridge, U.K., 1997). The data on the increase in obesity in the United States over the past century was taken from Lorens A. Helmchen's "Can Structural Change Explain the Rise of Obesity? A Look at the Last 100 Years," published as a discussion paper for the Population Research Center at the National Opinion Research Center and the University of Chicago. Information on the World Health Organization's estimates of global disease burdens in 2020 is available on the group's website, www.who.int.

### CHAPTER 4: DEMENTED

Leo Navratil's book on his early work with the artists at Maria Gugging was entitled *Schizophrenie und Kunst* in German (Deutsche Taschenbuch Verlag, Munich, 1965). The Haus der Kunstler website URL is www.gugging.org.

Tony Monaco and his colleagues' work on *FOXP2* was published in *Nature* 413:519–23 (2001). Svante Pääbo and his colleagues' work on *FOXP2* in the Vindija Neanderthal appeared in *Nature* 418:869–72 (2002). The involvement of the Mount Toba eruption in the population bottleneck that occurred around 70,000 years ago was proposed by Stanley Ambrose in the *Journal of Human Evolution* 34:623–51 (1998). Christopher Henshilwood and colleagues' work on the etched ochre from Blombos Cave was published in *Science* 295:1278–80 (2002). Richard Goldschmidt's theory on macromutations and "hopeful monsters" was described in his book *The Material Basis of Evolution* (1940, reissued by Yale University Press, New Haven, 1982). Richard Lenski and colleagues' computer model for the evolution of complex traits appeared in *Nature* 423:139–44 (2003). Richard Lewontin and Stephen Jay Gould's notion of evolutionary spandrels was described in the *Proceedings of the Royal Society of London, Series B* 205:581–98 (1979). Marshall Sahlins's reference to the "original affluent society" appeared in *Man the Hunter*, referenced above.

Toby Lester's article on the sounds of modern life, "Secondhand Music: The Chance Harmonies of Everyday Sounds May Mean More Than We Think," appeared in the April 1997 issue of the *Atlantic*. Clive Gamble's description of Upper Paleolithic warfare along the Nile appears on page 190 of his book *Timewalkers: The Prehistory of Global Colonization* (Harvard University Press, Cambridge, Mass., 1996). Robin Dunbar's analysis of neocortex ratio and group size appeared in *Behavioral and Brain Sciences* 16:681–735 (1993), from which Figure 24 was taken. Further discussion of the significance of the number appeared in Louise Barrett, Robin Dunbar, and John Lycett's textbook *Human Evolutionary Psychology* (Palgrave Macmillan, Basingstoke, U.K., 2001).

The data on antidepressants being the most widely prescribed drugs in the United States comes from the CDC (www.cdc.gov) and appeared in the publication *Health, United States, 2007*. It was widely reported, notably by CNN, when the report was released.

### CHAPTER 5: FAST-FORWARD

Most of the material in the Whitakers' story was taken from my interview with them, though some of the details were widely reported in 2002–03 on the BBC News website (news.bbc.co.uk). More information on Diamond-Blackfan anemia is available on the website www.diamond blackfan.org.uk.

The data on age at first birth in the United States was taken from the CDC report "National Vital Statistics Reports, Volume 51, Number 1: Mean Age of Mother, 1970–2000," and that on European mothers was taken from Eurostat (ec.europa.eu/eurostat). The data of IVF success rates was compiled by the CDC in its publication "2005 Assisted Reproductive Technology (ART) Report: National Summary," available on their website. The 2004 study on PGD and IVF implantation rates was by Magli et al., *Human Reproduction* 19:1163–69 (2004). The *Los Angeles*

*Times* article on PGD and IVF success rates was published in the April 27, 2007, edition. Eric Lander's quote appeared in an article in *The Scientist* published in January 2002. Ellen Ruppel Shell describes the research efforts on obesity in *The Hungry Gene* (Grove, New York, 2003).

The French court case in which a child with Down's syndrome sued for being born was widely reported in the press, for instance in *The Independent* (U.K.) on November 29, 2001. The survey showing that a majority of people would test for heart disease susceptibility was published in the *Journal of Genetic Counseling* 18:137–46 (2009). The new project in China tracking children after genetic testing was featured on the CNN news website (www.cnn.com) on August 5, 2009. The testing is being conducted by the Shanghai Biochip Corporation.

Geoffrey Wills's study of bebop musicians appeared in the *British Journal of Psychiatry* 183:255–59 (2003). Charles Limb and colleagues' MRI study on jazz musicians was published in the online scientific journal *PLoS One* (www.plosone.org) on February 27, 2008. Arnold Ludwig's *The Price of Greatness: Resolving the Creativity and Madness Controversy* (Guilford Press, New York, 1995) examines the relationship between creativity and mental illness. David Horrobin describes his theory of the relationship between schizophrenia and creativity in *The Madness of Adam and Eve* (Bantam, New York, 2002).

The ant/acacia study appeared in *Science* 319:192–95 (2008). Leon Kass wrote about "the wisdom of repugnance" in an article in the June 1997 (volume 216) issue of *The New Republic.*

### CHAPTER 6: HEATED ARGUMENT

Various global warming scenarios are reviewed in Mark Lynas's book *Six Degrees: Our Future on a Hotter Planet* (National Geographic, Washington, D.C., 2008). Tuvalu's threat to bring a lawsuit against the United States and Australia was widely reported in the media, particularly on the BBC News website. The auto sales data presented in Figure 28 was taken from Edmunds Inc., *The New York Times,* and J. D. Power and Associates.

The report of the Intergovernmental Panel on Climate Change, *Climate Change 2007: The Physical Science Basis* (*Summary for Policymakers*), is available at www.ipcc.ch. The controversy over the hockey stick graph is reviewed on the website http://www.worldclimatereport.com/index.php/2005/03/03/hockey-stick-1998-2005-rip/.

The 2008 Audubon Society Christmas bird count summary is available at http://www.audubon.org/bird/bacc/Species.html.

Bernice de Jong Boers's article on Mount Tambora appeared in the October 1995 issue of the journal *Indonesia* (volume 60). The Tambora story is told in *Volcano Weather: The Story of 1816, the Year Without a Summer,* by Henry Stommel and Elizabeth Stommel (Seven Seas Press, Newport, R.I., 1983). The cultural effects of the Tambora eruption were discussed in the January 29, 2005, issue of *New Scientist.*

*The World Disasters Report 2001,* published by the International Federation of Red Cross and Red Crescent Societies, estimated that more people were forced to leave their homes because of environmental disasters than war—around 25 million at that time, and the number has increased since then. Energy usage in America versus India was discussed in Thomas Homer-Dixon's excellent book on peak oil and the energy crisis, *The Upside of Down: Catastrophe, Creativity and the Renewal of Civilization* (Island Press, Washington, D.C., 2006). There are many good descriptions of the pebble bed reactor available on the Web. A good overview of desalination technologies is available in the September 2007 issue of *Scientific American,* as well as in an article by Kathryn Kranhold in the January 17, 2008, issue of *The Wall Street Journal.*

## CHAPTER 7: TOWARD A NEW MYTHOS

The prisoner's dilemma was first described in John von Neumann and Oskar Morgenstern's *Theory of Games and Economic Behavior* (1944, republished by Princeton University Press, Princeton, 2007). Lawrence H. Keeley's study on violence in "primitive" societies is detailed in his book *War Before Civilization: The Myth of the Peaceful Savage* (Oxford University Press, Oxford, 1996).

The lifetime chance of being murdered in the United States (1 in 100) was estimated in Michael Ghiglieri's book *The Dark Side of Man: Tracing the Origins of Violence* (Perseus Books, Reading, Mass., 1999). The Moriori and their extinction are discussed in Jared Diamond's *Guns, Germs, and Steel.* Karen Armstrong discusses Sayyid Qutb and the history of fundamentalism in her fascinating book *The Battle for God* (Ballantine, New York, 2001).

The survey detailing public acceptance of evolution around the world appeared in *Science* 313:765–66 (2006). The *Economist* article quoting

Stephen Ulph on Internet usage among Islamic fundamentalists is "A World Wide Web of Terror," in the July 12, 2007, issue. The estimate of the rates of language loss is from Daniel Nettel and Suzanne Romaine's book *Vanishing Voices: The Extinction of the World's Languages* (Oxford University Press, Oxford, 2002). The statistics on Facebook users can be found on Facebook's website (http://www.facebook.com/press/info.php?statistics), and the average number of users was discussed in a February 26, 2009, article in *The Economist.*

The following books were cited in the final section of this chapter: Bill McKibben, *Enough: Staying Human in an Engineered Age* (Times Books, New York, 2003); James Howard Kunstler, *The Geography of Nowhere: The Rise and Decline of America's Man-Made Landscapes* (Simon & Schuster, New York, 1993); and William McDonough, *Cradle to Cradle: Remaking the Way We Make Things* (North Point Press, New York, 2002). James Howard Kunstler's *Washington Post* editorial appeared in the May 25, 2008, issue of the newspaper.

# Index

## About the Author

SPENCER WELLS is an Explorer-in-Residence at the National Geographic Society and Frank H. T. Rhodes Class of '56 Professor at Cornell University. He leads the Genographic Project, which is collecting and analyzing hundreds of thousands of DNA samples from people around the world in order to decipher how our ancestors populated the planet. Wells received his Ph.D. from Harvard University and conducted postdoctoral work at Stanford and Oxford. He has written two other books, *The Journey of Man* and *Deep Ancestry*. He lives in Washington, D.C., with his wife, a documentary filmmaker.

## About the Type

This book was set in Garamond, a typeface designed by the French printer Jean Jannon. It is styled after Garamond's original models. The face is dignified, and is light but without fragile lines. The italic is modeled after a font of Granjon, which was probably cut in the middle of the sixteenth century.